Old and New Battlespaces

OLD AND NEW BATTLESPACES

Society, Military Power, and War

Jahara Matisek and
Buddhika Jayamaha

LYNNE
RIENNER
PUBLISHERS

BOULDER
LONDON

Published in the United States of America in 2022 by
Lynne Rienner Publishers, Inc.
1800 30th Street, Boulder, Colorado 80301
www.rienner.com

and in the United Kingdom by
Lynne Rienner Publishers, Inc.
Gray's Inn House, 127 Clerkenwell Road, London EC1 5DB
www.eurospanbookstore.com/rienner

© 2022 by Lynne Rienner Publishers, Inc. All rights reserved

Library of Congress Cataloging-in-Publication Data
Names: Matisek, Jahara, 1983– author. | Jayamaha, Buddhika, 1971– author.
Title: Old and new battlespaces : society, military power, and war / Jahara Matisek and Buddhika Jayamaha.
Other titles: Society, military power, and war
Description: Boulder : Lynne Rienner Publishers, Inc., [2022] | Includes bibliographical references and index. | Summary: "Explores how today's civil society, technology, and military organization are dramatically transforming the character and conduct of war"— Provided by publisher.
Identifiers: LCCN 2021028801 (print) | LCCN 2021028802 (ebook) | ISBN 9781626379961 (hardback) | ISBN 9781955055079 (ebook)
Subjects: LCSH: War. | Military science. | War and society. | Military policy—United States. | Strategy.
Classification: LCC U21.2 .M3644 2022 (print) | LCC U21.2 (ebook) | DDC 355.02—dc23
LC record available at https://lccn.loc.gov/2021028801
LC ebook record available at https://lccn.loc.gov/2021028802

British Cataloguing in Publication Data
A Cataloguing in Publication record for this book
is available from the British Library.

Printed and bound in the United States of America

∞ The paper used in this publication meets the requirements
of the American National Standard for Permanence of
Paper for Printed Library Materials Z39.48-1992.

5 4 3 2 1

Contents

Preface vii

1 Imagining War   1
2 How Battlespaces Change Through Time   17
3 Warfare in the Nuclear Age   45
4 The Complexity of Emerging Battlespaces   63
5 Compressed, Converged, and Expanded Battlespaces   97
6 Civil Society and the Contemporary Battlespace   117
7 New Battlespaces and Strategic Realities   141
8 The Future of Grand Strategy   159

List of Acronyms 173
Bibliography 175
Index 189
About the Book 197

Preface

THE GENESIS OF THIS BOOK LIES WITH OUR PERSONAL OBSERVAtions of warfare over the last two decades, alongside teaching warfare, strategy, and military studies at universities. Many student comments and questions pertain to the wars that they have grown up with in the twenty-first century. Their probing provided us with the inspiration and impetus to see this project through.

Students pose incisive and intuitive questions honed to all levels of war and all instruments of national power, such as the following: What is the desired political end state in Afghanistan? Are the political and military objectives there aligned? The United States has been continually engaged in Iraq for over thirty years, so what is the desired strategic end state? What strategic advantages have the United States derived as a nation after two decades of war? Is the digital community we call cyberspace vulnerable to attack? Who is supposed to defend the cyber domain, and how? What does automation mean for the future of airpower and the Joint Force broadly? How does the exponential growth of technology change the character of war? How do weaker adversaries outflank and defeat great powers and create strategic realities? What does war mean if everything from terabytes of data to vehicles and smart devices can be weaponized?

These questions focus on several themes. Inquisitive students of strategy want to understand the character of the wars that they may be called on to fight, how wars change over time, and how to properly conceive of a battlespace. For many military leaders, the traditional concept of their job entails the management of violence to varying degrees, and the battlespace denotes a leader's concept of the area of operation in order to employ combat power effectively. What is the nature of that battlespace, and how can one make sense of it across space, through well-known and emergent domains, and along the various levels of war? Where should one look, how should one look, and for what should one look?

The conversations that began in classrooms, then continued via text messages, emails, and video chats with colleagues at home and deployed abroad, transformed into this book during the Covid-19 pandemic. We aim to bridge the scholarship divide among academics, policymakers, and strategists who are interested in emerging security threats and who want to make better sense of a transformed US-led international order.

* * *

This book is based on work supported by the Air Force Office of Scientific Research under award number FA9550-20-1-0277 and was cleared for public release (PA#: USAFA-DF-2021-130). Any opinions, findings, and conclusions or recommendations expressed in this material are those of the authors and do not necessarily reflect the views of the United States Air Force.

We are indebted to the Department of Military and Strategic Studies and the Department of Research at the US Air Force Academy for supporting our endeavors during this book project.

1

Imagining War

WAR IS THE ORGANIZED USE OF VIOLENCE IN PURSUIT OF AN OBJECtive. The link between war and politics prominent in contemporary discourse emerged with the advent of standing armies at the end of the early modern period in Europe and particularly after the Napoleonic Wars. The purpose of military strategy is to reconcile ends, ways, and means to shape the behavior of the adversary. Violence and the credible threat of violence are merely means to create a particular effect; destruction is an intermediate effect, never an end in and of itself.[1]

War is also an elemental social enterprise. As societies change, so too do the ways they organize violence—such as the shift from feudal levies to national conscription to maintenance of standing armies. As societies change, the nature of the objective changes, sometimes quite drastically—such as from territorial conquest to the defense of human rights. Likewise, the character of warfare evolves as societies change. How societies adopt varied forms of military organizations shapes the character of war just as war shapes the character of societies.[2] When trying to understand military power and the origins of and victors in a conflict, it is widely accepted that the character and shape of war is influenced by economic and political power and other structural variables.[3] There also is a general sense that broad trends, like the intensification and diversification of global economic ties, the development of new technologies, the increasing wealth of many countries, and the political mobilization of their peoples, has affected the relationship between the organized use of violence and the creation of effects that have become contingent and far more complex than in the past.

If military history is a guide, the relationships between violence and creating effects will get more complex over time. This reality poses a Janus-faced puzzle for students and practitioners of military and strategic studies. Students of strategy usually look backward to understand changes in warfare over time. Ultimately, strategists have the "task of turning one currency—military (or economic, or diplomatic, and so forth) power—into quite another (desired political consequences)."[4]

Practitioners who design strategy attempt to apply lessons of the past as they address what they think are salient ongoing changes. Ultimately, they execute strategy in an environment of imperfect information with uncertainty as the only certainty. Standing athwart the past and the present, students of strategy are faced with concerns over what broad societal attributes will contribute to the changing character of war over time. But how should both students and practitioners define the character of warfare across time and space? What aspect of war, such as, for example, its nature, character, representation, interpretation, or organization, changes over time? We develop the concept of battlespaces as the prism with which to see changes in war over time.

Napoleon's Grande Armée marched across Europe in a context that was distinct to the kinds of societies and states that existed at the dawn of the nineteenth century. Napoleon was able to harness a commitment and commonality of purpose that came with a nascent French national identity. The armies that he faced lacked this resource. Union generals in the American Civil War prevailed over their Confederate adversaries in a context undergoing rapid change as industrialization shaped society and state power in the middle of that century. Mechanized infantry and strategic bombing of cities from World War II have become iconic images of conventional warfare.[5]

The dominant mid-twentieth-century concept of warfare focused on the clash of armies of sovereign states that build and sustain complex military organizations. That kind of warfare, particularly viewed from the present, seemed to pursue clear objectives and offered little ambiguity concerning the main protagonists. Other types of warfare, deemed "small wars" in a 1940 US Marine Corps manual, were confined to the geopolitical periphery or were sideshows to bigger wars.[6] But as the information age changes societies and economies, organizational principles change and adapt to various efficiencies.[7]

Taken together, these broad changes reflect different configurations of power. Material resources are important throughout, though in varied ways. Broad political developments, such as the rise of nationalism, the value of global connectivity, and evolving global norms about individual and group rights also weigh heavily in how war is imagined. So too are changes in how governments believe they can effectively organize and project military power.[8]

The post–World War II era saw the increasing prominence of conflicts referred to as irregular warfare.[9] These conflicts included nonstate armed groups (i.e., "insurgents") fighting colonial and indigenous state forces to win what these insurgents defined as national liberation. They fought adversaries that had far greater resources, including firepower. But the insurgents in many cases had backing from like-minded supporters around the world. Irregular warfare spawned all kinds of terms—such as *unconventional war, guerrilla warfare, counterinsurgency, counterterrorism*, and so forth—and fit in the broader category of civil wars fought within internationally recognized boundaries.[10] Civil wars (i.e., intrastate), then and now, require a different analytical framework for discerning the logics of violence, compared to the more familiar experience of interstate wars that dominated the popular and practitioner imaginations well after the experience of World War II.[11]

Recent and ongoing conflicts in places like Afghanistan, Iraq, Libya, and Yemen present images of ambiguity. They seem to have no clear beginning or end, and distinctions between civilians and combatants, though hardly unique to this time and these places, can appear especially blurred. Remotely operated weapons such as drones complicate the picture. Mercenaries appear to play a growing role in many of these conflicts. Though these elements of warfare are not necessarily new, their combination at this point in the global configuration of power and the roles states play in people's lives leave Americans with difficulty saying what these wars are about and whether the United States has won or lost in places like Iraq.[12] They wonder why American soldiers are deployed to so many countries that they cannot find on a map.[13] Even some in Congress were surprised to find that US forces were in Niger. This only came to their attention when four Green Berets were killed in an incident in October 2017.[14] Beyond the United States, there are many armed factions fighting in myriad cities, towns, and villages in Libya (2014–present), Syria (2011–present), Somalia (1980s–present), Myanmar (Burma) (1948–present), and across the Sahel of Africa (1958–present). But this violence is applied deliberately in pursuit of an objective, as has been the case for these sorts of old battlespaces in recorded human history.[15] In irregular wars, unlike in conventional wars, one is forced to look carefully to discern clarity in the logics of violence.[16]

Warfare and Contours of Battlespaces

In the definitional landscape, *war* refers to the event in its totality. *Warfare* is the sum total of means utilized. Within it are analytical distinctions and typologies such as *conventional, unconventional, regular, irregular, symmetric,* and *asymmetric*. A *battlespace* in a war is both a cognitive reality and a manifest reality. A battlespace as a *cognitive reality* is a combat leader's understanding of the area of operation with the aim of using combat power effectively. A battlespace as a *manifest reality* is the many processes in the area of operation that violence affects, as well as processes that affect violence—the endogeneity (and often confusion) of violence inherent in any battlespace. In war, a combat leader is a manager of violence; hence, coercion is projected toward creating a specific effect that in its cumulative manifestation contributes toward realizing a specific objective. The level of complexity and the many aspects of a battlespace, not to mention understanding it, are also contingent on the level of war—that is, the positionality of the combat leader in the broader war at the strategic, operational, and/or tactical levels.

The supreme commander Napoleon saw his battlespace in terms of the strategic level—though his genius lay in his understanding of the battlespace along all levels of war and also in how well he aligned political and military end states.[17] He knew the broad objectives, the desired effects his grand army was expected to create, forcing adversaries to sue for peace. This is because war is more than just a simple battlefield. Napoleon's efforts at making a French country made his war-fighting efforts more effective because he reshaped the post-Revolutionary French state. The way he mobilized the French nation and the military for war shaped both the state and the wars he fought. Meanwhile, Napoleon's grand marshals, at the operational level, were aware of their commander's intent, maneuvering military elements under their command accordingly.[18]

Combat leaders at the tactical level conceive their battlespace in terms of enemy movements. Combat leaders, then as now, operate in an environment of imperfect information and employ combat power under their command understanding that they must minimize their own vulnerabilities while exploiting enemy weaknesses. A leader at the strategic level may not know the fine-grained details of the battlespace at the tactical level, just as a leader at the tactical level may not know the strategic-level details and challenges. But once war commences, the idea is that, on the one hand, the desired military end state is aligned with the desired political end state, and, on the other, the strategic, operational, and tactical levels remain aligned and work in lockstep. Understanding a battlespace is an enterprise predicated on conditions of uncertainty and imperfect information.

In this old battlespace conception, at the strategic level Napoleon conceived of the battlespace in terms of his desired end state—getting the enemy to sue for peace—and in terms of the enemy's and his own military capabilities. He would then maneuver his armies toward a military dénouement with the enemy forces—his military genius being his attention to detail at all levels and his astute use of geographic depth and space seamlessly with his military formations. It was war fought in the single domain of land and to a lesser degree on the seas, and no one doubted that this was war. Not all wars are perceived as clear-cut today; debates rage about the blurring of war, conflict, and competition.[19] However, we do know that there are myriad ways to create effects other than waging a war. What would the conception of a battlespace be for tomorrow's military leaders?

Complexity of Contemporary Battlespaces

Hostile activities in Ukraine and in the South China Sea provide explicit examples of the challenges of military actions and violence in the twenty-first century. The Russians outthought and outfought their competitors for power in the Black Sea region, creating a civil war and managing to expand territory and alter national borders with the use of violence. Similarly, the Chinese created a new imagined reality in the South China Sea—namely, that this region was inherently owned by China. The historical and geopolitical significance of each action, not to mention the dominant narratives and themes generated about each region, challenges notions of an American-led order.

In early 2014, "little green men" (Russian *Spetnaz* commandos) appeared in Crimea, and with explicit use of violence for the first time since World War II with no pretentions to the contrary, Russia acquired new territory and made it part of its country.[20] Simultaneously, Russia created proxy forces to support a civil war in Ukraine, where none existed previously. By the time the rest of the world recognized what was transpiring, Russia had realized its objectives, and the rest were left to deal with a new reality. One is hard put to point out whether a war was involved in the way Russia acquired Crimea. Yet, within days of these unmarked Russian commandos occupying Crimea, Wikipedia labeled the region as Russian, and *National Geographic* added Crimea to Russia, with the lead map maker stating, "We map de facto, in other words we map the world as it is, not as people would like it to be."[21] Russia realized objectives with subtle, deceptive, and effective use of implicit and explicit violence. The internet quickly rewarded this behavior by socially and digitally constructing the reality Russia sought. The success of Russia dominating the information environment even led to a top American general admitting that Russian actions in Ukraine were "the most amazing information warfare blitzkrieg we have ever seen."[22]

While the South China Sea saw numerous regional claims to the area beginning in the 1970s, China began building artificial islands in 2013 and employing numerous naval militias (dressed up as fishermen) to impose a new reality on the sea. This tactic continues with large flotillas that take seemingly permanent shelter on tidal atolls from bad weather that does not materialize. Meanwhile, the world was deliberating whether these acts were a provocation, an act of war, or what not.[23] The net outcome is an altered geopolitical reality that China's neighbors must navigate. Are these actions acts of war? Can China build artificial islands in parts of the ocean without a legal mandate to do so? Do the atolls and oceanic space China claim become their area of control despite international legal condemnation?

These examples—a random sample out of multitudes of human conflicts—capture instances in which the protagonists relied on either the explicit use of violence or the implicit threat of violence to generate effects. Moreover, in each instance, a military leader as a manager of violence can be informed by history but is forced to reckon with a battlespace at variance from ones that came previously, as warfare continuously evolves. That raises the broader puzzle examined in this book: What does it mean when one refers to the changing nature of battlespaces, and what shapes them across time and space?

Throughout history, land and then the sea were the domains where the fortunes of armies were decided. In the early twentieth century, the air domain became an integral part of war fighting, followed by the space domain at the end of the century. At the turn of the twenty-first century, the cyber domain has become increasingly vital to defend and maintain. Its prominence and importance in day-to-day activities has led to many debates about a coming era of cyberwarfare.[24] What has become abundantly clear from these discussions is that the cyber domain will be a complementary domain in pursuit of political objectives.[25] Previously, many technology alarmists had engaged in "FUD" (fear, uncertainty, and doubt) debates to create the perception that the main line of effort in modern warfare would be primarily waged via cyberspace.[26] Regardless, the importance of cyber conflict brought in the blurring of statecraft and warfare and a blending of all domains and jointness along the spectrum of cooperation, competition, and conflict.

Technology-induced change inadvertently created a humanly devised domain, conceived as cyberspace. This humanly devised domain is in turn anchored in outer space, and it is not an exaggeration to suggest that nearly every individual and modern institution on the planet is dependent on the space domain and the satellites that orbit Earth. From daily life to fighting wars, the on-demand connectivity afforded by cyberspace, the Internet of Things, and satellites has integrated domains in ways inconceivable previously. Consequently, besides the domains of land, sea, and air, the cyber and space domains have become emergent war-fighting battlespaces—if not direct, then indirect areas of competition—since activities in the traditional domains of land, sea, and air are becoming dependent on cyber and space.

For example, the 2011 Libyan War became the first conflict in which 100 percent of the munitions used were precision-guided weapons, meaning that the ability of the North Atlantic Treaty Organization (NATO)–led coalition to hit targets was dependent on data-connection links afforded by the cyber and space domains.[27] However, this continual historical trend toward more precise standoff weapons (i.e., the ability to project violence without being close to an enemy) is only part of the change.[28] Use of violence and war is not about destruction; it is about creating effects, with necessary destruction as an intermediate effect. However, there are alternative avenues to create sociopolitical effects in the domination of the information environment without the use of violence in generating real strategic outcomes.

State and nonstate actors engage in malicious cyber activities inside the United States and allied nations with impunity, such as the remarkable 2020 SolarWinds Hack that resulted in the compromise of sensitive data of major firms like Microsoft and numerous US government agencies.[29] These cyber actions are taken against corporations, state entities, and even specific individuals. American citizens subjected to disinformation, misinformation, intimidation, bullying, or threats from an external state and/or state-affiliated proxy in cyberspace could call the local police, and yet the police would be incapable of defending them.[30] This raises another set of questions: Do such actions rise to the level of war? If not, what does it mean if the state cannot protect its citizenry from foreign aggression? Is the country failing in its social contract? What does this tell us about the present and future when the idea of "reaching out and touching someone" has assumed new meaning due to advances in technology?

The traditional image of what constitutes conventional war and what does not is predicated on violence. This old battlespace conception is no longer so clear-cut in liberal democracies where people do not live inside digital fortresses as citizens do in China, Russia, and other authoritarian countries. In the "free world," civil society is vulnerable in a way that was simply not possible before the advent of global connectivity and instantaneous communications. Various effects and end states are pursued in the cyber domain that achieve a slow, steady erosion of Western values and ideational foundations. This is an important distinction of sociopolitical-information warfare: it differs from traditional notions of information warfare or political warfare in that it achieves nonkinetic effects of "anomie" by trying to break social order and solidarity.[31] It is a distinct paradigm shift in which society and varying social dimensions are put under constant duress without adversaries making explicit use of violence.[32]

For example, in 2014 anti-vax movements were amplified and gained prominence due to specific online tactics. This included a snowballing effect, especially on Twitter with bots and the creation of private Facebook pages, adding more followers that would reshare antivaccine posts and amplify polarizing views against the basics of fundamental vaccine science.[33] This has real-world consequences: a 2015 study found that failure to vaccinate in the United States resulted in almost $15 billion in economic damage.[34] China and Russia view this as an opportunity to weaken their adversaries. Operatives in these countries have waged cyber campaigns of disinformation about vaccines, especially in the Covid-19 era, not to mention about the origins of certain diseases (e.g., Soviet Union's 1980s Operation Infektion, blaming the United States for inventing HIV/AIDS).[35] Hence, the societies in which individuals have open access to information have become centers of gravity in battles to influence attitudes that question the legitimacy of national authorities and even their political systems—what used to be called *subversion*, a term seldom used in policy circles since the end of the Cold War. These information operations are cheap to execute and present few risks, particularly if executed by authoritarian states that do not allow foreigners the same level of access to their own citizens' information environment.

Tangible violence, with all the risks and costs associated with pursuing certain political and military end states, appears to be waning as a primary line of effort. Gaining importance in these new battlespaces is the growing emphasis on pursuing nonkinetic efforts to generate effects that achieve strategic objectives with less need for armed personnel physically present in a location—although there will always be utility in having someone on the ground with a gun.[36] Many of the risks and costs associated with old battlespaces seem to have decreased in importance with the new battlespaces developing. However, the thread between all battlespaces is that along the spectrum of competition and conflict, warfare essentially remains a human activity.[37]

External actors increasingly target Western civil society by leveraging new domains (i.e., cyber and space), successfully implementing sociopolitical-information warfare against a society by creating rifts and exploiting grievances.[38] Much as rebellious colonials angered British Redcoats by not fighting the proper eighteenth-century way, twenty-first-century adversaries are circumventing traditional notions of military power and strength by attacking and exploiting an undefendable position: civil society.[39] In line with historical precedents in which the weak exploit power asymmetries, these emerging competitors and spoilers seek ways of diminishing the inherent strengths of the strong by shifting toward a strategy with less emphasis on kinetic military options. This fundamental shift requires deciphering what can be weaponized, including intangible things like data to target individuals anywhere in the world and achieve political and military objectives. Hence, we are forced to consider what value an airstrike in the twenty-first century has when adversaries can attack Western civil society via online media—that is, by exploiting the sociopolitical-information environment—which can be as persuasive and impactful in establishing certain narratives in a target country.[40]

As great power competition (GPC) became a defining point of the national security strategy laid out by the Donald Trump administration in 2017, the US military and many of its allies continue to focus on large-scale combat operations (LSCO) to fight the next war in ways they would prefer if they must fight. Meanwhile, anti-Western actors invest in cheaper asymmetric, non-LSCO capabilities, hoping to achieve gains without a kinetic fight. The GPC era (2017–present) of a rising China and resurgent Russia is a by-product of America's failure to develop a proper strategic vision since the end of the Cold War. Moreover, it is an extension of the growing tension in the international system due to countries like China and Russia wanting to maximize their own autonomy and survivability. Their warfare pursuits have been to achieve gains without provoking a military response from the West.

And what is war if adversaries create tactical and strategic effects without the use of violence? This elemental reality makes civil society in liberal democracies an emergent battlespace. But the necessary legal structures designed to safeguard civil society prevent security professionals from even placing it within the broader discussion of emergent battlespaces. Yet the unavoidable reality is, any young lieutenant who steps into a war and attempts to understand a specific battlespace inevitably faces a series of integrated domains in land, sea, air, cyber, space, and civil society. These domains, while analytically autonomous, remain integrated in reality, and the interconnectedness creates an emergent reality. The question then is, Do all these changes constitute a fundamental change in war fighting, and how can one make sense of these changes?

About the Book

This book turns war into the unit of analysis. Seeing war over time means seeking to understand the changing face of battlespaces through time. We provide a temporal continuum that runs right across the discussion. Turning war into the unit of analysis and change in war over time into the outcome raises the analytical issue of the level of analysis. If change in war over time is the outcome in need of explication, is the level of analysis situated at the strategic, operational, or tactical level of war?

This book develops a macro-level interdisciplinary framework drawing on military and strategic studies, political science, sociology, history, and even literature to delineate and discuss war over time. Simultaneously, it uses derivatives of the macro-level categories to delineate war along the levels of analysis. The word *delineate* is used decidedly. The aim here is to derive concepts and categories from the broad framework with which to delineate contours of change across time and along the levels of war. This subsequently means seeing what war would mean if adversaries could create effects either without the use of violence or with innovative use of violence.

This analytical approach also makes this book different from the many recent books that discuss the changes in warfare, with some explicitly speaking of a fundamental change in war. Most of them isolate adversaries' cyber activities without explicitly discussing them in terms of civil society, much less seeing them in terms of the emerging multidomain environment. This book is informed by some of their technological discussions and ideas of international relations, yet differs from them by turning warfare into the unit of analysis.[41]

The chapters in this book proceed as follows:

In Chapter 2, we develop the overall analytical framework. War is an elemental human enterprise. War is also a paradoxical enterprise that brings out the best and worst in humans, a nature that ensures war remains one of the most studied topics. If there is a consensus about war, it is that it is inherently complex. Such complexity—usually referred to in terms of fog and friction—comes from the reality that the usual ideational, economic, military, and sociopolitical processes that shape people every day begin to change at a rapid pace as a function of violence. These changes in turn also shape the dynamics of violence, and the process becomes iterative, creating a level of endogeneity where parsing out precise causal processes becomes extremely difficult. War fighters in this milieu are the managers and executors of violence. They make decisions on the application of combat power based on their specific understanding of the battlespace, where they both inhabit and shape the processes.

This analytical framework takes war in its totality with all its complexity. It is based on the premise that the changing *character* of war over time is best understood and viewed through the relationship that lies in the dichotomy between the nature of war and the character of war. In the annals of military and strategic studies, the nature of war is seen as constant over time due to the immutable purposive nature of human beings. It is therefore assumed that since the nature of war remains constant, the character of war, the bloody manifestations of violence, is what changes over time. If the nature of war is constant over time and the character of war changes over time, what explains the changes over time?

We build on the long-held assumption that the nature of war is constant. Waging war is in the nature of human beings as purposive individuals. Therefore, wars are always fought for a purpose. Based on that assumption, the elemental premise of this book argues that the ultimate character of war is defined by the immutable social sources that shape the exercise of organized violence and the contingent decisions of human beings in the way they apply violence. People organized into a community, tribe, clan, city-state, empire, nation, and/or nation-state can choose to wage war in pursuit of an objective. The nature of the objective sought and the social sources of military power—geopolitics, regime type, ideas, nature of military organizations, and scientific knowledge (GRINS)—ultimately shape the character of war over time and across space.

Our GRINS framework builds on classics that discuss war but differs in its manifestation by making the relationship between the nature of warfare and the character of warfare the locus of analysis. The premise is that understanding what shapes the relationship between the nature and character of war provides the guideposts that allow one to delineate the broad contours of the character of warfare over time. The rest of the chapter takes the long view of history and covers familiar ground. It interprets the familiar, from the Napoleonic Wars to the end of the World War II, with a novel lens in the hope that this allows readers to see the familiar differently. On the temporal dimension, this chapter concludes at the end of World War II as nuclear weapons gain prominence.

In Chapter 3, we discuss the utility of warfare as a means of generating effects in the presence of nuclear weapons. The advent of nuclear weapons was a defining moment in human history. The development of this weapons system had far-reaching strategic ramifications and played a decisive role in shaping warfare and geopolitical realities. It paradoxically enabled power while inhibiting freedom of statecraft.

War has long been a means of realizing sovereign objectives, ranging from building states where none existed to expanding territory and building empires. Yet in an age of nuclear weapons, when friends and foes alike possess the capacity to scorch the earth, what is the role of warfare? States, like individuals, are purposive, animated by interests, concerns, and fears. The organized capacity for violence will therefore always be one of the many instruments of national power, but how shall one wield it in creating effects, in pursuit of objectives, if it will mean collective annihilation?

As the nature of war is constant, people will always find a way to use violence as a means. Nuclear weapons do not make war obsolete. But they do force considerable thought about innovative ways to exercise violence in pursuit of objectives. This chapter discusses how warfare evolved during the Cold War as nuclear weapons made "interstitial warfare" the new normal.[42] Specifically, forms of social power—ideational, economic, military, and political—get formally or informally institutionalized. Domestic institutions, international organizations, transnational alliances, and so forth, are manifestations of institutionalized power arrangements.[43] Once in place, these arrangements shape the behavior of individuals, communities, and even states. Interstitial warfare refers to conflict that takes place either at the edges of or between institutionalized power arrangements and often makes prolific use of proxies to avoid direct confrontation among nuclear-armed adversaries.

During the Cold War, the geopolitical distribution of power shaped the rise of this interstitial warfare. Political warfare played a role in the ongoing struggle between the United States and the Soviet Union to shape political outcomes in each other's sphere of influence. Though rudimentary from a contemporary perspective, these efforts raise important considerations and provide guidelines for thinking about cyberwarfare and influence operations today. Proxy warfare was another interstitial operation, with considerations that are important to consider when reflecting on potential futures of warfare in a more competitive geopolitical environment. The chapter concludes with the implications of the end of the Cold War.

In Chapter 4, we discuss warfare during and after the end of the Cold War. The transformative moment was when the Berlin Wall fell, and there was no forceful response from Moscow. The end of Soviet influence led to an immense concentration of ideational, economic, military, and political power in the hands of the United States and its allies. Rapid political and economic change around the world altered economic incentives. Technological innovations took place at a rapid pace, changing the way people conducted their daily interactions. Moments of abrupt change also led to social dislocations, opening new political opportunities to settle scores or create new realities.

While some intractable wars ended, new ones emerged. The concentration of power was such that decisions made in the United States, the European Union (EU), and NATO played a decisive role in shaping wars from the end of the Cold War to the beginning of the twenty-first century. Since the 9/11 attacks, NATO and the United States have been fighting wars for twenty years, and for over thirty years if one counts the imposed no-fly zone in Iraq and numerous military operations in the fractious Balkans. We refer to this permissive environment as the unipolar moment, a time in which US and NATO partner politicians and their foreign policy establishments engaged in what we call "strategic narcissism" (with due nods to Hans Morgenthau and H. R. McMaster), as this geopolitical moment seduced many into assuming that the world could be remade into the Cold War victors' image and that the underlying nature of warfare no longer applied.

This chapter concludes with how the subsequent decisions of adversarial states, nonstate actors, and assorted spoilers also began to play a role in shaping the character of wars. It ends with a discussion of the slow decline of Western strategic primacy in the traditional domains and how the West was unprepared for the rise of emergent domains.

In Chapter 5, we discuss the rise of battlespaces as multidomain realities. We especially focus on what makes the contemporary reality different and how the confluence of more competitive geopolitical realities and technological changes has integrated traditional war-fighting domains of land, air, and sea with the cyber and space domains. Simultaneously, these changes have also created novel battlespaces—or social realities that adversaries can (and do) leverage into battlespaces.

The chapter examines how liberal democracies find themselves at a distinct disadvantage in this altered strategic and war-fighting context, whereas adversaries have become proficient at generating strategic realities with tactical maneuvers. Picking up on the theme of expanding battlespaces woven throughout this book, the chapter takes up the issue of "lawfare," of adversaries' disingenuous use of the legal principles of liberal democracies and of major international organizations to achieve alternative political ends.

Chapter 6 dives deeper into the nature and reality of integrated domains and the nature of emergent domains, especially implications for warfare being less kinetic and lethal, yet becoming increasingly more effective. It elaborates the contrasting strategic conceptions of the victors of the Cold War and the rest, as it were. Then it discusses the logic of how revisionist powers utilize emergent domains to outthink, outsmart, and outfight the United States to create strategic realities.

This chapter focuses on how adversaries with authoritarian political systems exploit new opportunities to engage in interstitial warfare by weaponizing the open societies of countries that have liberal democratic systems. Recalling some of the conclusions from Chapter 3's attention to political warfare, this chapter highlights how technological changes and the effects of strategic narcissism in the United States and elsewhere contribute to the asymmetry of this form of contention.

The chapter builds on this element of interstitial warfare to better conceptualize cyberwarfare in the broader contemporary battlespace. It draws important distinctions between the use of cyber technologies in the pursuit of espionage (the collection of information for political and military purposes) and subversion (the transmission of information for the purposes of dividing and weakening an adversary from within). Though the latter is easily conceived as a political term that reflects the preferences and values of the observer, much like terrorism, there remain important distinctions in the ends, ways, and means of these uses of information.

Chapter 7 surveys contemporary strategic realities. We emphasize the expansion of the battlespace to new domains. US policy makers and planners recognize this expansion while also struggling with how to adjust bureaucracies and policies to reflect these changes and at the same time preserve commitments to the basic values of an open society. Previous chapters showed how similar challenges were addressed in the past. Legacy institutions of the Cold War military and the unipolar moment become obstacles to crafting flexible responses to contemporary challenges.

Finally, Chapter 8 concludes by posing a series of questions pertaining to the future of grand strategy in an era of integrated and emergent domains. It discusses the implications for Western civil society and the sort of adaptations needed to excel in the era of a new battlespace.

Notes

1. Burke, Fowler, and McCaskey, *Military Strategy, Joint Operations, and Airpower.*
2. Downing, *The Military Revolution and Political Change.*
3. For example, Morgenthau, *Politics Among Nations*; Beckley, "Economic Development and Military Effectiveness."
4. Gray, *The Strategy Bridge*, 7.
5. According to US doctrine, the dichotomy for military war fighting is organized between traditional and irregular warfare. For the purposes of this book, the "conventional warfare" phrase is utilized in lieu of the "traditional warfare" term, as traditional warfare has substantially different meaning for anthropologists studying indigenous conflicts. For more on this doctrinal discussion, refer to Joint Publication 1, *Doctrine for the Armed Forces of the United States* (Washington, DC: Joint Chiefs of Staff, 2017), I-5; Ferguson, "Masculinity and War."
6. US Marine Corps, *Small Wars Manual*, 1940, Internet Archive, https://archive.org/details/UsmcSmallWarsManual1940Reprinted1990/mode/2up.
7. Davidson and Rees-Mogg, *The Sovereign Individual.*
8. Beckley, "The Power of Nations."
9. This growing prominence of irregular warfare was a function of the growing number of civil wars due to the Cold War era of two hegemons trying to avoid nuclear annihilation, so supporting proxies (i.e., rebels) was a by-product of their great power competition. This resulted in supporting weaker forces that then fought the much stronger polity. However, civil wars are a common phenomenon since ancient Rome. Armitage, *Civil Wars.*
10. According to US doctrine, unconventional war is considered a subset of irregular warfare because it translates into working with militias in a denied area to fight against the regime and/or occupying military powers. Bunker, "Unconventional Warfare Philosophers."
11. Joint Publication 1.
12. "NEW POLL: Americans Want Troops Home from Afghanistan, Iraq; Opposed to More Military Engagement," Charles Koch Institute, January 23, 2020.
13. Bijal P. Trivedi, "Survey Reveals Geographic Illiteracy," *National Geographic*, November 20, 2002.
14. Loren DeJonge Schulman, "Working Case Study: Congress's Oversight of the Tongo Tongo, Niger, Ambush," Center for a New American Security, October 15, 2020.
15. Thucydides, *History of the Peloponnesian War.*
16. Joint Publication 1.
17. Gibbs, *Military Career of Napoleon the Great.*
18. Durham, *The Command and Control of the Grande Armée.*
19. Nadia Schadlow, "Peace and War: The Space Between," *War on the Rocks*, August 18, 2014; Hoffman, "Examining Complex Forms of Conflict."
20. Ryan Faith, "The Russian Soldier Captured in Crimea May Not Be Russian, a Soldier, or Captured," *Vice News*, March 10, 2014.
21. Brian Resnick and National Journal, "Should Wikipedia Put Crimea on the Russian Map?," *The Atlantic*, March 19, 2014.
22. Gen. Philip Breedlove quoted in John Vandiver, "SACEUR: Allies Must Prepare for Russia 'Hybrid War,'" *Stars and Stripes*, September 4, 2014.
23. Mackubin Thomas Owens, Bradley Bowman, and Andrew Gabel, "Dangerous Waters: Responding to China's Maritime Provocations in the South China Sea," *National Interest*, December 20, 2019.

24. Healy, *A Fierce Domain*.
25. Rid, *Cyber War Will Not Take Place*.
26. Perlroth, *This Is How They Tell Me the World Ends*.
27. Mueller, *Precision and Purpose*, 4.
28. O'Connell, *Of Arms and Men*.
29. Brian Barrett, "Russia's SolarWinds Hack Is a Historic Mess," *Wired*, December 19, 2020.
30. For the purposes of the book, the nuanced difference between misinformation and disinformation is intent. Thus, disinformation results from intentional disingenuous behavior to spread false information for malicious purposes, while misinformation results from the unwitting spread of bad information.
31. The problem of anomie (i.e., breakdown of societal values, behaviors, and norms) was first identified by Émile Durkheim in 1893. Durkheim, *The Division of Labour in Society*.
32. Levite and Shimshoni, "The Strategic Challenge of Society-centric Warfare."
33. Renée DiResta and Gilad Lotanscience, "Anti-Vaxxers Are Using Twitter to Manipulate a Vaccine Bill," *Wired*, June 8, 2015.
34. Ozawa, "Modeling the Economic Burden of Adult Vaccine-Preventable Diseases in the United States."
35. Broniatowski, "Weaponized Health Communication"; Carmen Paun and Susannah Luthi, "What China's Vax Trolling Adds Up To," *Politico*, January 28, 2021.
36. Wylie, *Military Strategy*, 85.
37. Storr, *The Human Face of War*.
38. Some might be tempted to interpret civil society as the "human domain." However, definitions of this domain, first outlined by US Special Operations Command in 2013, are narrowly detailed as "the totality of the physical, cognitive, cultural, and social environments that influence human behavior." US Special Operations Command, "Human Domain White Paper, Version 7.5," MacDill Air Force Base, Florida, May 2013, 4–5. For more on the human domain and its conceptual issues requiring integration of the cyber domain, see Gregg, "The Human Domain and Influence Operations in the 21st Century."
39. Hibbert, *Redcoats and Rebels*.
40. Through mentorship of an undergrad student, Christina Durham came up with the "thought bombs" analogy. Christina Durham, "Thought Bombs of the 21st Century: Memes as a Tool in Influence Operations," PIPS Research Paper, College of William & Mary, May 2021.
41. For example, Singer and Brooking, *LikeWar*; Patrikarakos, *War in 140 Characters*; Watts, *Messing with the Enemy*; Stenge, *Information Wars*; Pomerantsev, *This Is Not Propaganda*; Benkler, Faris, and Roberts, *Network Propaganda*; Brose, *The Kill Chain*; Zuboff, *The Age of Surveillance Capitalism*; Clarke and Knake, *The Fifth Domain*; Rid, *Active Measures*; Buchanan, *The Hacker and the State*; Howard, *Lie Machines*; Kreps, *Social Media and International Relations*.
42. Our conception of interstitial war is different from the first description of Clausewitzian interstitial politics in this 2014 article: Craig, "Intermediarized Security Governance and the 'Sultans' Retort.'"
43. Mann, *The Sources of Social Power*, 1:26.

2
How Battlespaces Change Through Time

EVERY WAR IS PARTICULAR IN ITS OWN WAY. OF THE MANY SMALL and large details that shape a war, human ingenuity plays a defining role in how warfare evolves over time. Each war in history, irrespective of the objective, has generated tactical, operational, and strategic innovations.[1] Though scientific knowledge plays a significant role in shaping the character of a war, how militaries and armed actors incorporate applied scientific knowledge in terms of technology fundamentally shapes how actors view and operate in battlespaces. For example, Richard Gatling invented a rapid-fire multibarrel firearm during the American Civil War, setting off a long stream of innovation in automatic light weapons. He thought this more efficient weapon would reduce the size of armies, which would mean that fewer people would be killed in combat.[2] His Gatling gun saw limited use by Union forces that adapted the weapon to light river patrol craft. Meanwhile, European armies incorporated the weapon into campaigns of imperial conquest. The more advanced Maxim gun defied Gatling's prediction of reduced casualties, and in Sudan at the Battle of Omdurman in 1898, an army commanded by the British general Sir Herbert Kitchener reported casualty figures as follows: "20 Britons, 20 of their Egyptian allies, and 11000 Dervishes lay dead."[3]

This is just one example of technological advances in warfare. Some of these advances have tactical implications just as much as they have strategic implications. The greater challenge is in understanding how the incorporation of these advances drives change in war over time. It means asking the following: What shapes contingent choices of protagonists, driving change over time? Where should one look to assess change? Trying to understand changes in warfare over time allows one the chance to probe a pressing question of our time: Are we entering a new era of warfare with battlespaces unlike those of previous eras? We anchor our assessment in existing literature that builds on the nature-versus-character-of-war dichotomy to explain change in terms of the changing character of warfare.

Nature Versus Character of Warfare

If there is a central axiom in the study of war and strategy, at least in the Western cannon, it is the claim propounded by Carl von Clausewitz (1780–1831) that the *nature* of war remains constant, while the *character* of war is always changing. Consequently, any change in war is perceived as changing the character of war. In the Clausewitzian view, war is a duel on a grander scale.

> War is thus an act of force to compel our enemy to do our will. . . . Force—that is physical force, for moral force has no existence save as expressed in the state and the law—is thus the *means* of war; to impose our will on the enemy is the object. To secure that *object* we must render the enemy powerless; and that in theory is the true aim of warfare.[4]

The form—that is, the character—is dependent on time and the place. War, defined most broadly as imposing one's will on another by force, is shaped by the context in which it plays out. Therefore, the *character* of warfare is constantly changing. But the *nature* of war remains constant since human nature is constant over time. Human beings are purposive and goal driven, with pursuits to maximize power, pleasure, or whatever objective, and in those pursuits, violence is seen as a means to an end. The contention here is that to understand what defines the direct relationship between the nature and the character of war is to be able to capture change over time.[5] Though war is replete with complex, interdependent, and cross-cutting processes of violence—a mechanistic conception of war for purposes of clarity—one can imagine the *nature of war* as an independent variable and the *character of war* as a dependent variable (i.e., the outcome), with numerous intervening variables that define the relationship between nature and character shaping the character of warfare.

The distinction between the nature and character warfare is more than semantic. Its utility comes from knowing that the nature of war is constant. Its nature is unchanging because of how people have consistently behaved over several millennia. As violence has been a part of the enduring nature of human existence, war's character changes in line with contextual factors (e.g., laws, culture, politics, societies, power relations between states, etc.).[6] For instance, Frank Hoffman notes that "the accelerating introduction of cyber weapons, robots, and artificial intelligence will be major sources of change in how war is conducted."[7] His observation points to how these changes will alter relationships within Clausewitz's trinity, shaping war and society—and vice versa.

On War and Clausewitz

Clausewitz conceptualized war in its totality and also along its levels: policy, grand strategy, military strategy, operations, and tactics. That is one of the many reasons why his writing has had a more enduring effect in Western war colleges than that of his contemporary Antoine-Henri Jomini (1779–1869), who focused more on the operational and tactical levels, providing a prescriptive approach.[8] *On War* seeks to answer a series of puzzles borne out of his experience: specifically, What is war? How does one make sense of war? And do any timeless determinants shape warfare?

Clausewitz was a product of the Enlightenment. *On War* captures and synthesizes all the ideational tensions of *his time*. The Prussian military of which he was a part was itself an incipient product of novel, rationalist thinking, organized by Frederick the Great (1712–1786). This monarch was highly influenced by (and became a patron of) Enlightenment thinkers. He called Niccolò Machiavelli "the enemy of mankind" and then assiduously followed his advice to behave instrumentally in state building—perhaps at the behest of his intellectual mentor and another product of the Enlightenment, Voltaire.[9] Frederick the Great thus pursued a pragmatic approach to building new state institutions in Prussia that would buttress his power. A formidable, well-organized, and disciplined army with a professional officer corps was among his innovations. Napoleon introduced his own distinctive form of efficiency in harnessing military firepower to a mass army that he employed with great effect along the strategic, operational, and tactical levels of war. When Napoleon marched into Prussia, he made a point to stop at the grave of Frederick the Great to pay homage, saying, "Hats off, gentlemen, if he were still alive, we should not be here."[10] Napoleon mobilized this revolutionary élan to destroy the old regimes of Europe and establish French hegemony over much of the Continent.

Clausewitz wrote *On War* at the height of Enlightenment thinking. His work fits with attempts to develop grand theories with universal validity. He constructed a theory of war while trying to make sense of war. Unlike some of his enlightened contemporaries who maintained more absolute faith in the predictive powers of reason, Clausewitz—the experienced soldier—knew better. Clausewitz understood Murphy's law by experience: no plan survives first contact with the enemy. More importantly, he also experienced how bureaucracies organized according to rationalist logic could be put to use in realizing war objectives in previously unthinkable ways. He understood that reason has its moment in war just as much as irrationality and chance. Therefore, his writing—an inductive theorizing enterprise—strikes a balance between Enlightenment and Counter-Enlightenment thinking. One can see in his *On War* the influences of Enlightenment thinkers such as Francis Bacon, René Descartes, and Isaac Newton, alongside early Romantics like François-Auguste-René de Chateaubriand and Johann Gottlieb Fichte, who were later seen as Counter-Enlightenment thinkers.[11] Indeed, Clausewitz eventually strikes a balance to capture the ideational tension that animated most Enlightenment and Counter-Enlightenment thinkers of the time in his conception of the paradoxical trinity—a concept that has informed generations of military thinkers.

The Paradoxical Trinity

The paradoxical trinity is the classic expression of the Clausewitzian ontological position on war. It speaks to the reality of war as

> composed of primordial violence, hatred, and enmity, which are to be regarded as a blind natural force; of the play of chance and probability within which the creative spirit is free to roam; and of its element of subordination, as an instrument of policy, which makes it subject to reason alone.[12]

In present day terms, the trinity consists of the *people* (passion), the *government* (policy), and the *armed forces* (chance). Yet Clausewitz asserts that the nature of war remains constant over time since his theoretical scaffolding is built on claims of natural law and natural rights. The most defining feature is that there is an intrinsic inalienable human nature that remains immutable over time.

On War has since become military orthodoxy, and scholars diligently abide by the nature-versus-character dichotomy. As a result, studying war over time has meant studying the changing character of war and rarely the relationship between the nature and character of war. We put the relationship at the center of our analytical framework.

Military power is "socially organized, concentrated lethal violence."[13] War is the utilization of concentrated lethal violence in the form of military power in pursuit of an objective, usually cast in political terms but possibly to gain ideological, economic, or political power. Whatever the objective sought, in war there is a relationship between the objective, character, and outcome. The many elements that are believed to shape the character of war are in fact what we term the social sources of military power that generate the endogenous processes that eventually shape warfare in its totality.

Nature, Character, and Outcome in War

War is a social enterprise. Though violence—kinetic power—is seen to be defining in war, our premise is that social sources of military power ultimately shape how wars evolve over time.

A turning point in human evolution was the discovery of the inherent power an object possesses due to its motion—namely, the force of kinetic energy. Moreover, if kinetic energy is projected properly, it can be used to bend others to one's will. Other generations of humans likely discovered, through creative and destructive processes, the organizational power behind their group if they are able to swing—literally—the most coordinated amount of kinetic energy in terms of socially organized concentrated lethal violence.[14] This was the formative process that informed the creation of military power. Whether harnessing kinetic power from stones, bludgeons, varieties of metal projected from catapults, explosive charges, or energy emitted with the use of chemical reactions, war is still seen to be predicated on force turned into violence for the purpose of bending an adversary to one's will. Ever since the discovery that military power helps in the pursuit of collective objectives, warfare has been a social enterprise.[15]

Social sources of military power—analytically distinct yet related in reality—in varied combinations shape its ultimate evolution. These, in varied combinations, shape the character that we attribute to specific wars in terms of conjuring imagery of war, old and new. Social sources of military power that shape the character of war are geopolitical context, regime type, ideas, nature of military organizations, and scientific knowledge (GRINS). These five variables, which can best be mentally compartmentalized as GRINS in Figure 2.1, provide guideposts for how different

Figure 2.1 The Nature and Character of Warfare as Viewed Using the GRINS Framework

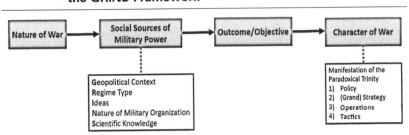

processes have evolved in warfare, shaping the relationships between nature, character, and outcome. It is a framework for understanding how these factors shape and influence the changes in war over time.

The GRINS model is a way of getting past old battlespace notions of military power and effectiveness, which were ideas best put forth by Allan R. Millett, Williamson Murray, and Kenneth H. Watman in 1986.[16] Later in 1988, Millet and Murray provided much deeper assessments of military organizations with a highly influential three-volume book series in 1988.[17] These contributions made scholars reconsider the way power and national strength were exercised across the political, strategic, operational, and tactical levels of war, each impacting military institutions in varying ways and translating into differing levels of performance. Then in 1995, Stephen Rosen simplified these variables by explaining how society and social structures—and the ways in which they impact military organizations—influence a country projecting power.[18]

Further building on the notions of what contributes to military effectiveness, Risa Brooks and Elizabeth Stanley acknowledged how important the global environment was in impacting countries in their organization of war, specifically the ways in which culture, social structure, and political institutions interact in the processes of generating military power.[19] The GRINS model advances these lanes of scholarship by accommodating the strategic, operational, and tactical levels and their implications for old and new battlespaces, specifically with regard to how the character of warfare is shaped. Moreover, the GRINS model contributes to the growing literature on military power and effectiveness being more than just an emphasis on mass and war matériel, such as techniques of force employment in modern warfare due to increasing levels of lethality.[20] This suggests the utility of the GRINS model, up and down the different levels of war, as government capability and productive capacity are important at the strategic level but competency at the tactical level requires flexible and adaptable institutions. This is not to say that there cannot be more dimensions in the operationalization of strategy. Each level of war contributes to technological, operational, logistical, and social dimensions, which Michael Howard viewed as dominating the Western way of thinking about warfare in 1979.[21]

War in an Idealized World

People, being purposive, utilize military power in pursuit of an objective. Application of military power commences when the objective is identified, and the political and military end states are decided upon to match this guidance. When the war tocsin sounds, one attempts to discern the casus belli ("an act or event that provokes or is used to justify war") in terms of the objectives. The social sources of military power shape the overall character of war.

In an ideal world, political and military end states, the objectives sought with the use of violence, are captured in the policy guidance that sets military power in motion. Upon the onset of violence, assuming the policy level—the political and military end states—remains constant, there is a set of interactive dynamics between the three levels of war: (grand) strategy, operations, and tactics. The same social sources of military power, shaped by the objective, also affect the character of war along the levels of war.[22] That is, *if no war is ever the same*, one can understand the variations over time and along the levels of war, taking the context into account, by seeing the war in terms of the social sources of military power. By applying GRINS (Figure 2.1 above) as a way of understanding the relationship between the nature and character of war, one can find why this relationship matters in the outcome. And *if in fact different geographic regions in the same war hardly appear the same*, then one can understand the variations in the character of war by using the social sources of military power as the prism. This translates into focusing on a specific level of war or along the levels of war—that is, contextualizing the GRINS framework at a specific time and a place.

Understanding War: The Longue Durée

Nature of war: The nature of war remains constant on the premise that from the first days of organized human violence to the contemporary use of force, the nature of humanity remains constant. Humans are goal oriented, purposive, make contingent choices, but the nature of their goals, means of their pursuit, and their choices remain context specific. That does not mean individuals pursuing warfare are necessarily rational actors in the traditional sense—which itself a highly debatable point of contention.[23]

Objectives and outcomes: The objectives of the sovereign are shaped through time and space. Wars are fought in pursuit of an objective, and to realize the objective is to generate an outcome. War aims also evolve over time. Pursuing a war to create a world that could mirror the heavens with the use of violence is now inconceivable in the middle of modern-day Germany. But, for instance, the Thirty Years' War (1618–1648) that resulted in Germany's losing over one-third of its population was fought in pursuit of that objective.[24] Through much of history it was accepted that a sovereign in one country would invade a neighbor to create empires, contingent on limitations forced on them by the social sources of military power of the time. But with the defeat of Nazi Germany and the Japanese Empire in 1945, the prohibition of conquest became a cornerstone of the postwar international order. This norm, along with growing acceptance of the concept of self-determination, made possible the recognition of new states that had very limited capacities to maintain internal order.[25] This means that, as in Somalia from 1991, even the total collapse of a central government does not lead to the extinguishing of that state's identity as an actor in the international system. These changes in the nature of the international system of states play a big role in the appearance of categories such as "failed states" and the proliferation of persistent conflicts that are fought in and around failed states.

This is an example of the sorts of endogenous processes that shape warfare. Human nature may be constant, but human desires and objectives are shaped by the time and the place—that is, ideational foundations influence what is acceptable and what is not. The same ideational foundations—naturally, there is never one ideational foundation but multiple contending currents, some hegemonic, some not, and how social power is organized plays a role—form aspects of the social sources of military power just as they shape human objectives and desires.

For example, waging a war in pursuit of European-wide hegemony today is no longer a strong tool for social mobilization in European countries and would meet a decidedly more vigorous response in targets of such a campaign, perhaps among the many contemporary states that were founded as components of European empires and that retain international boundaries inherited from their colonial pasts. The process of endogeneity lies in the reality that the ideational foundations that shape desired outcomes of leaders and people become an ideational scaffolding that shapes warfare. Nature is constant, but desires vary, and turning desires into outcomes, with varying degrees of possibility, is ultimately shaped by the social sources of military power that, when war commences, manifest in the paradoxical trinity—character traits that are generalized as defining a war.

Social Sources Supporting Military Power

Military power is the socially organized, concentrated use of lethal violence. It derives as an emergent need, a form of power, and, as best described by Michael Mann,

> from the necessity for organized defense and the utility of aggression. Military power has both intensive and extensive aspects, for it concerns intense organization to preserve life and inflict death and can also organize many people over large socio-spatial area.[26]

The nature of military power is shaped by social sources in varied combinations. Thus, as the next section will show, breaking down warfare and battlespaces by the GRINS approach enables a more robust contextual understanding of the relationship between the nature, character, and outcome of war.

Geopolitical Context

Neighborhood matters. The historical and contemporary reality is that since the beginnings of organized societies and with various regimes emerging, every polity, irrespective of the type, has had to seriously comprehend its respective neighborhood—the geopolitical context that it inhabits. Geopolitics captures the relationship between regimes and the human geography of territorially bounded people—the many societal power relationships among and between them. Geopolitics can cause war and influences all wars, whether wars between regimes or warfare among territorially bounded people, now referred to as civil war. But how such war takes place is contextually contingent on time and place.

In the long view of history, one certainty has emerged. War—or the threat of war—has been a constant means of altering territorial boundaries or changing regimes and/or has served as an alternative form of diplomacy. The certainty rulers enjoy today, of territorial integrity and state sovereignty, is a historical anomaly in the *longue durée* that emerged only as a deliberate political construct at the end of World War II and was codified in the charter of the United Nations. There is no certainty that this precept will remain constant. This could change, and there are limited but important signs that this may be happening. Russia, for example, annexed Crimea in 2014 after seizing it from Ukraine by force, and Israeli control of Palestinian territories since 1967 appears to be more widely accepted within the Middle East than was true for many decades in the past. Much of this order appears to depend on how status quo powers interact with up-and-coming challengers and on how regional balances of power change.[27] In simple terms, the neighborhood where one lives always matters, and how is a question of what one desires and the allotted time.

Regime Type

A regime captures how power is organized among a geographically bounded people, or society, over time and across space. According to sociologist Michael Mann's epistemic position, a particular regime represents how power is organized in a society.[28] Regimes define, govern, and shape human interactions among each other and with those in power. Regimes range across history from clans to tribes, kingdoms, city-states, theocracies, empires, and states (e.g., authoritarian, democratic, etc.) and—if history is any guide—are bound to emerge in innovative and hybrid forms for the foreseeable future. Since the nature of the regime both defines and shapes how people interact with each other and how people navigate power relationships, regime type heavily influences and shapes the character of war.

Regimes are formed by agglomerations of four sources of social power: ideological, economic, military, and political power. Though the power sources remain constant across time and space, their character can change, and regime survival is dependent upon adaptation. For example, during the process of modern state formation in Europe, most prosperous city-states were unable to adapt to the structural changes in Europe and were absorbed into states.[29] Similarly, the Ottoman Empire struggled to adapt to the internal and external challenges it faced. When Ottoman leaders attempted to adapt via a modernization plan, the imperial structure was too brittle, and modernization only hastened its decline.[30] The great variation of regime types across time and space—tribes, clans, empires, states, and so forth—is best explained by how power sources reform and crystalize in varied combinations, forming specific regime types.[31]

Ideas

Ideas constitute perhaps the most nebulous concept in the discussion of evolving war. "Ideas," as stipulated here, remain distinct from scientific knowledge, which constitutes a specific body of commonly accepted ideas. Ideas are a vague analytical concept because they are whatever what one thinks they are, unless one is told otherwise. This is the epitome of a social construct in that ideas achieve value, with attributes of fact, based on collective agreement. Without proper boundary conditions, ideas can be a concept that seemingly explains everything but explains nothing accurately. Ideas are also nowhere to be seen, and yet one begins to see them everywhere when one recognizes the features with which to identify them.

Religions constitute bodies of ideas that attempt to build a bridge between the temporal and the celestial, between the known and the unknown. Religions attempt to capture realities that surpass experiences. Religions, in building a bridge between the known and unknowable, also generate specific codes of behavior and moral and ethical precepts with which people relate to each other. According to Michael Mann,

> Shared understandings of how people should act morally in their relations with each other, are necessary for sustained social cooperation. Durkheim demonstrated that shared normative understandings are required for stable, efficient social cooperation, and that ideological movements like religions are often the bearers of these.[32]

Ideas via religion speak to the opposite of the rationality that resonates in each human being; they are variously referred to as beliefs and convictions, expressed as "faith." Ideas, purely in the temporal dimension, turned into ideological movements, can have the same effect as religions. For example, Marxism, national socialism (i.e., Nazism), and other forms of contextually dependent nationalist movements demonstrate how secular ideational movements can have similar effects as a religion. For example, while sometimes temporary and fleeting, there is in nationalism a captivating aspect that surpasses sensory experience, making it possible for people to fervently commit to and die for an intangible idea: the nation.

In addition, "ideas" in the temporal dimension also constitute humanly devised concepts or frames of references with which people attach meaning to sensory perceptions, make sense of and understand the past and present, and attempt to predict the future. They also shape interactions in terms of social norms and traditions and sometimes form the basis of formal and informal institutions. When ideas are formalized and institutionally captured, they provide a formal basis for human interactions.[33] For example, American citizens take for granted the idea that "we hold these truths to be self-evident, that all men are created equal, that they are endowed by their Creator with certain unalienable rights, that among these are Life, Liberty, and the pursuit of Happiness."[34] This social construction of what should be obvious truths did not always exist. In fact, there is nothing self-evident about it, as it was hypothesized to be self-evident, a post-Renaissance ideational construct, based on Enlightenment thinkers' assumptions about natural rights and natural laws. Having been promulgated, it is imposed institutionally as self-evident, turning an idea into a manifest reality, and it has been so well internalized that it is indeed self-evident.

Ideas ultimately have a form of power over and among people that is at once ephemeral and ethereal and also concrete and manifest. Politics and policy are ideas. Sir Isaiah Berlin wrote, "It may be that, without the pressure of social forces, political ideas are stillborn; what is certain is that these forces, unless they clothe themselves in ideas, remain blind and undirected."[35] Unlike anything else, ideas have a restraining and liberating effect on people and are the only social source that can acquire a sense of momentum by making themselves irresistible to multitudes. To quote Isaiah Berlin again, "Over a hundred years ago, the German poet Heine warned the French not to underestimate the power of ideas: philosophical concepts nurtured in the stillness of a professor's study could destroy a civilization."[36] He further notes that Heine

> spoke of Kant's *Critique of Pure Reason* as the sword with which European deism had been decapitated, and described the works of Rousseau as the bloodstained weapon which, in the hands of Robespierre, had destroyed the old régime; and prophesied that the romantic faith of Fichte and Schelling would one day be turned, with terrible effect, by their fanatical German followers, against the liberal culture of the West.[37]

Ideas have a defining role in shaping how people live and what they do; they consequently define the times and ways in which people live. It is the value and power of ideas that turns time into epochs of history.

Ideas can also shape both why wars are fought and how they are fought. For example, complete annihilation of a group of people (i.e., genocide), elimination of those who hold different ideas (i.e., politicide), and the forcible expulsion of millions of people (i.e., ethnic cleansing) are not acceptable military end states in the twenty-first century.[38] But they were acceptable—though highly contested—war aims not too long ago. That is a direct impact of altered ideational foundations. Similarly, the simple military reality of contemporary wars is that "civilian" casualties and collateral damage generate substantial attention and that deliberate targeting of civilians constitutes a "war crime." The ideas of war crimes, human rights, international law, and so forth, are all specific derivatives of broader ideational currents that play a role in shaping how wars are waged today.

Nature of Military Organization

Military organizations are a social source, just as they are part of and shaped by the society they come from. Often, but not always, military organizations are embedded in society, even as they maintain a level of autonomy from society. A military is socially embedded in the sense that its members represent the wider society and share similar norms and ideational foundations. Equally important is that just as much as the state can shape the military, the military can also shape the nature of the regime and the type of political institutions formed.[39] Military culture can also shape the coordination of activities at the strategic, operational, and tactical levels of war—and biases and assumptions can hinder performance across these levels when counteracting an adversary.[40]

A military can be relatively autonomous from society in the sense that a military garrison in a far-flung town (again, contingent on the nature of the military) is not affected by the local societal trials and travails, and the same military has the capacity to turn its arms against people in quick order. At the same time, an army can have a militarizing effect on society if not kept in check by civilian oversight.[41] Ultimately, a military organization has external and internal logics that shape its nature, such as what a military organization is, its organization, its purpose, and its internal objective and subjective functional logics.

A military organization is an agglomeration of people, ideas, and knowledge. There are specific organizational logics that shape a military. Its purpose, conventionally, is seen as the defense of the realm and, if necessary, offense in defense of or in expansion of the realm. But the notion of offense and defense is contingent on time and place.

The realm did not always mean people and territory, as it is understood today, especially in the West. It meant the ruler, and before militaries became the sophisticated organizations of today, they were the primary means of instilling internal rule of law and order. Machiavelli's dictum used to be the long animating principle, the raison d'être for organizing militaries: "the main foundation of every state, new states as well as ancient or composite ones, are good laws and good arms; and because you cannot have good laws without good arms, where there are good arms, good laws inevitably follow."[42]

Over time, as ideas evolved, regime types governing societies evolved. Since World War II mainly, militaries have hardly been seen as the primary means of instilling domestic law and order—at least in the West.[43] Logic of military organizations has become, from king and country, to different variations of duty, honor, country, but this is not always the case in the twenty-first century. For example, the People's Liberation Army of China is simultaneously the armed forces of the country and the militia of the ruling party, and when necessary, it will shoot China's own citizens in defense of the Chinese Communist Party's rule. Something similar is at work with the Iranian Revolutionary Guard Corps, whose aim is to defend the revolution, the regime and its elites, and, by extension, the nation. This is an example of how a particular combination of regime type and ideational foundations shapes the organizing logic and nature of a military organization. Regardless, military organizations and their respective cultures can dominate the way in which they coordinate activities at the strategic, operational, and tactical levels—not to mention the way they interact with political leadership.[44]

Scientific Knowledge

Scientific knowledge constitutes an accumulated body of knowledge derived through systematic investigations over time. Scientific knowledge in its pure form makes determinant discoveries, and these discoveries are true universally, as suggested by the old natural law from the Vincentian Canon: "Quod ubique, quod semper, quod ab omnibus creditum est" (That Faith which has been believed everywhere, always, by all).[45]

Scientific knowledge helps people understand the world, if proper discoveries are made, outside superstition. It drives the quest to know why the sky is blue and why the apple fell downward and not sideways. Scientific knowledge in its applied form is technology. At the same time, this can translate into national and military productive capacity, as applied scientific knowledge should function at the strategic, operational, and tactical levels in terms of orchestrating and coordinating complicated military organizations, such as an aviation service or the nuclear enterprise.

Given the rapid pace at which applied scientific knowledge (i.e., technological innovation) changes, it has become convenient to focus on applied science and to hypothesize technology as fundamentally altering the nature and character of war. As will become clear, scientific knowledge shapes how wars are fought, but the impact of scientific knowledge on the overall conduct of war always remains contingent, because how military organizations incorporate new technology and adapt is going to determine the effectiveness of new technology. Further, a basic level of scientific knowledge in a society—the level of development of a society, as it were—also matters in the adoption of technology in warfare.

Applying the GRINS Framework

The GRINS framework (as shown in Figure 2.2) provides a broad macro-level analytical framework to glean broad outlines and contours of the character of a particular conflict. Analytically distinct components of the framework in reality are related. Each component plays a role in shaping the character of a war, just as the relationships between them play a role, adding to the endogenous processes inherent in conflict. We illustrate with some broad examples below.

Napoleonic Wars: The broad contours of the Napoleonic Wars were defined by ideas animating the French Revolution, the post-Revolutionary French Republic, and the nature of the military organization that grew out of the First Republic (1792–1804) until Napoleon declared an empire. Napoleon in turn transformed the French state, building on the institutional scaffolding of the First Republic.

World Wars I and II: In both world wars, the onset of violence (objective of the war) can be attributed to regime types, ideational foundations of national glory (i.e., nationalism), and how the behavior of regimes was shaped by the specific geopolitical context of the time. However, in both instances the two wars were clashes of industrial organizations predicated on full-scale mass mobilization of national resources and resolve. Consequently, consistent with the myriad endogenous processes that shape the character of a war once violence commences, regime type, nature of military organization, and rapid

Figure 2.2 War and the GRINS Framework

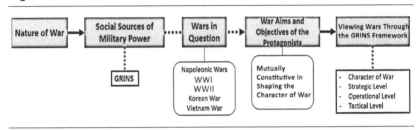

incorporation of applied scientific knowledge played a defining role in shaping the character of both wars.

Korean War: In every war, all elements play a role, while some can be defining. In the case of the Korean War, holding constant the unification objectives of the Democratic People's Republic of Korea (North Korea), the geopolitical context and regime type became defining. Once violence commenced, geopolitical concerns and ideational contestation of the time played a defining role at the strategic level, while the nature of military organization played a role in shaping operational- and tactical-level realities in the subsequent war of attrition. It would be mostly accurate to state that there was parity in terms of the incorporation of applied scientific knowledge—since both sides relied on legacy technology that dominated World War II.

Vietnam War: While the Vietnam War represented a complicated mixture of foreign intervention, proxy war, war of national liberation, and war of national unification, its transformation from a war of national liberation into what it became had everything to do with geopolitics and ideational contestations of the time. Once violence commenced, how each protagonist focused the efforts of its respective military organizations (strategies and tactics) and how each incorporated scientific knowledge with its tactics, techniques, and procedures became the public face of the war.

Social Sources of Military Power and Warfare over Time

The brief discussion below illustrates how the GRINS framework helps one understand changes in warfare and battlespaces over time by considering familiar conflicts: the French Revolutionary Wars, World War

How Battlespaces Change Through Time   33

I, and World War II. To repeat, the analytical premise is that geopolitics, regime type, ideas, nature of the military, and scientific knowledge are the social sources linking the nature of war, desired objectives of war, and what ultimately shapes the particular manifestation of the paradoxical trinity across time and space.

Wars of French Revolution

Seeing the French Revolutionary Wars through the macro-categories of the GRINS framework, one can see how ideas, regime type, and the nature of military organization played a defining role. Though ideas, regime type, and the nature of military organization are analytically distinct, they are inextricably related in practice. Each played a role, both on its own and in the way they became related as events unfolded, in shaping the character of war. Revolutionary ideas mobilized the people, and ideas in combination with the nature of the regime allowed for national conscription. Conscription, in combination with the modernized (in relation to the rest of the armies of the time) nature of military organization, was defining.

The French Revolution (1789–1799) upended existing social structures and reshaped power relations within France. However, the revolutionary spirit that pulsed within France had external implications. The geopolitics of the time—relationships between ruling entities, subentities, and human geography—was characterized by entrenched and consolidated absolutist monarchies of varying sizes and shapes and multi-ethnic empires on the Continent, with a series of mini-kingdoms scattered in between. Spain, Austria, and Russia constituted the largest land powers, and the Ottoman Empire bordered the Habsburgs.

The French Revolution, while a domestic enterprise, quickly generated geopolitical dilemmas. Revolution-induced social change within France made the conservative monarchical neighbors nervous, for they knew any spillover of revolutionary *ideas* would contribute to political upheaval. Distrust of neighbors was heightened by France's claim that its own animating revolutionary political ideas with universalist claims were fundamentally at odds with the governing principles of its neighbors—the divine right of kings.[46] Austria and Prussia attempted regime change to reinstate the deposed emperor, in their view the rightful heir to the French throne. In so doing, they hoped to return Europe to the tranquility that the old regimes had long enjoyed. It only worked to exacerbate the revolutionary fervor inside France. French revolutionaries countered with a historical first unthinkable to any of the kings claiming divine right. The French people defended the state and went on the offensive by building a military machine with the use of universal conscription, then overran the Low Countries (the Netherlands and Belgium) and, in quick order, created a patchwork of alliances. This shattered the geopolitical equilibrium that had shaped European politics.[47]

Napoleon took command as a junior officer in Valence in 1792, at a time when neither the French nor their opponents imagined the extent to which their actions would reshape Europe. Napoleon sought to translate universal revolutionary ideas into manifest reality by force of arms and with a distinct French imprimatur. All power corrupts, but some must grudgingly govern, and in that case, Napoleon would scramble to the top of the heap, personifying the Revolution. He managed to tear down the old order, though not to completely extinguish it, but then found it impossible to invent an alternative order when revolutionary claims had outspent the French material base due to military overreach.

The post-Revolutionary republic in France and death of the old regime was a necessary condition that made the French military juggernaut, personified by Napoleon, possible. The *regime type*—how ideological, economic, military, and political power was agglomerated into the republic—shaped the emergence and the life of the French military machine. By the time Napoleon rose as the undisputed leader, it was military power that defined the organization of the French state. The French military defined and created the state, which in turn defined the conduct of war, with the French nation a by-product, as Eugen Weber argued in *Peasants into Frenchmen*.[48] Indeed, influential scholars identify this endogenous process as leading to the creation of states—at least until the end of World War II. As Charles Tilly famously wrote, "War made the state, and the state made war."[49] He argued that interstate warfare explains the radical reduction of the number of states in early modern Europe and the increasing capabilities across many dimensions beyond the conduct of war of those that survived periods of intense warfare.

Other power sources, ideological, economic, and political, became subservient to military power. Yet no matter how intensive and sociospatially diffused the authority of the French military, military objectives would not have had traction in the minds of people and resonance in their hearts without the ideational power that animated the revolutionary spirit. *Ideas* shaped the Napoleonic Wars alongside innovations in military organization and technological developments in weaponry. The same idea that animated the French—and the notion of being French was a recent creation—also resonated in the hearts and minds of people abroad. This helped France in generating support for its enterprise in conquered lands. The aspiration to turn the stultified European status quo on its head—liberty, equality, and fraternity, with its explicit anticlerical appeal—inspired people.[50] It was also the beginning of the idea and the reality of the nation-state, the power of which became clearer during World War I. In pursuit of military objectives, military planners were not circumscribed by moral and ethical precepts. The desired end states seemed to justify the means—as was the mood of the time—in itself a particular ideational consensus.

The *nature of the military organization* was unique to the time. The external organizing logic was national, with universalistic pretensions, which allowed Napoleon's Grand Army to incorporate militaries of conquered territories into its own. Internally, the Grand Army, with its innovative corps organizing logic, provided a level of autonomy to the grand marshals, who shaped how massive elements of people and steel—military elements—could be efficiently maneuvered, as lumbering as they were by today's standards.

Human ingenuity always matters, but how human ingenuity plays a tactical, operational, and strategic role is contingent on whether the organization allows for such autonomous thinking. The French Republic—with an internal logic of its own—became an unstoppable force of military and ideational power. It could project its military and ideational strength outward, becoming an empire, subjugating and incorporating its neighbors, only in a particular geopolitical context. Sea power played a role, but the decisive domain remained land.

Scientific knowledge in its pure and applied form played a defining but not a decisive role—that is, with the scientific revolution in full swing, France incorporated rational precepts into its military organization, just as it incorporated applied scientific knowledge in terms of advances in technology into the military organization. This brief overview raises the question of ultimate primacy: Which, of the five sources, played the defining role in shaping war? Naturally, it is a combination of all five, but in this instance, it would be a safe infer that *ideas*—revolutionary upheaval—became the casus belli. But side by side, regime type, geopolitical context, nature of military organization, and scientific knowledge shaped the eventual war. The revolutionary ideas were a necessary condition. An animating theme was the destruction of the ancien régime, and the Congress of Vienna, sometime after, attempted to stitch together a semblance of balance of power predicated on conservative regimes.

World War I

If not for the losses involved, World War I would be in the annals of history as the greatest parody ever conceived. In any war, there is usually a casus belli that provides a tangible reason for declaring war. In this instance, an assassinated heir to the Austrian throne led the Austrian leadership to declare war on Serbia, though it would take Austria nearly until the end of the war to actually get into the fight. By then the logic of alliances and military mobilizations meant France, Germany, and Russia had fought themselves into a standstill. Seen broadly through the GRINS framework, geopolitical uncertainties, the relationship between the regime types, and the nature of the military organizations played a defining role in providing the unending supply of human fodder in a fight for ideas of nationalism. These factors were followed by how technologies were integrated into military organizations. The same broadly could be said of World War II, but regime type and technologies played a more defining role in shaping the character of that war.

World War I marked the collapse of a geopolitical balance in continental Europe that had existed in various forms since the end of the Napoleonic Wars. Altered geopolitical realities changed the balance of power between nations and empires in continental Europe. Prussia, long the main battlefield of Europe and the weakest of the great powers, harnessed rapid industrialization and a growing sense of shared identity among increasingly urbanized speakers of the German language to politically transform itself into a formidable land power in the form of Germany. Europe was divided along two opposing blocs. Germany, Austria-Hungary, Bulgaria, and the Ottoman Empire on one side as the Central Powers opposed Great Britain, France, Russia, Italy, Romania, Japan, and eventually the United States as the Allied Powers.

Soldiers in the millions from all sides were thrown into a relentless "storm of steel" generated by their respective adversaries, pitting mass armies against one another.[51] The mechanized battlefield produced by an agglomeration of new technology generated casualties that were not anticipated by professionals of violence. In the first year alone (1914), there were over a million killed and an additional five million casualties. A year and a half later, Germany and France lost half a million soldiers each during the Battle of Verdun. For France, that loss was about 2.5 percent of the country's entire male population.

Though advances in technology meant aspects of the war took place in all domains, it was on land where most deaths occurred. The fact that many military leaders had not come to terms with technological advances in weaponry made their battlespaces particularly lethal. Many of them clung to the notion that victory could be obtained through the maneuver of their military forces to fight decisive battles to smash opposing armies. But the advent of accurate longer-range rifles and battlefield innovations such as the use of poison gas, coupled with the mobilization of politically committed mass armies, meant deadlock.[52]

The power of nationalism to sustain a war of attrition to this extent was a novelty in the history of warfare up to that moment. Napoleon succeeded in mobilizing a mass army, but the geopolitical environment and technology of that time meant that he applied that force in campaigns of maneuver against relatively weaker adversaries. Though the staggering numbers of casualties eventually prompted political resistance among many citizens, this war of attrition wore on. Nationalist ideas and mass mobilization harkened back to the innovations of Napoleon, but this very different context fundamentally changed the roles that these features played in the character of warfare. The addition of US troops from 1917 and growing domestic political instability helped end the stalemate, leading to Germany's surrender in 1918.

World War I had a decisive impact that the GRINS framework helps to illuminate. In geopolitical terms, the United States' decisive role in breaking the military stalemate in its allies' favor established it as an Atlantic power. But the United States did not play the defining role in ending the war, only a decisive role. During the Paris Peace Conference, at which the victors debated key provisions of the postwar settlement, American president Woodrow Wilson attempted to negotiate a broad rules-based settlement that lacked American public support.[53] But the idea of self-determination of people at the center of Wilson's liberal worldview had a decisive influence. His claim that peoples in defeated countries had the right to national self-determination illustrates the integrated role of ideas in the GRINS framework. Bloody stalemate undermined the legitimacy of empires such as Russia and Austria-Hungary. The former hosted a socialist revolution (as Germany nearly did), and the empire's breakup produced the independent, ethnically defined states of Poland, Lithuania, Latvia, Estonia, and, for a brief time, Ukraine, Georgia, and Armenia. Likewise, out of the collapse of the Austro-Hungarian Empire arose Austria, Hungary, Czechoslovakia, and Yugoslavia.

Technological advances ("scientific knowledge" in GRINS terms) caused some like Brig. Gen. William ("Billy") Mitchell to think seriously about air power as a means to break deadlocks of land-based armies in future wars. Italian general Giulio Douhet also emerged as an air-power theorist. He was a key proponent of strategic bombing of cities in aerial warfare as an efficient way to break the will of an adversary's citizenry to continue fighting.[54] Mitchell and Douhet both conceived of air power as a means to save lives through the restoration of the possibility that an armed force could deliver a decisive blow to the adversary's capacity to wage war. Their approach expanded the definition of the battlespace to consider more direct warfare against the economy and the political system of the adversary. B. H. Liddell Hart's proposals for mechanized warfare in the form of motorized mobile forces was offered as another solution to the defensive stalemate, drawing from the introduction of tanks on the battlefield later in the war to support infantry.[55] His aim to reintroduce maneuver through the use of motorized artillery came to fruition when incorporated into Nazi Germany's concept of blitzkrieg in World War II. Liddell Hart thought that a professional, nonconscript mobile army would avoid the concentration of masses of soldiers that defined how battles were fought in World War I.

World War I introduced levels of state institutional coordination that expanded the concept of military organization within the GRINS framework. In the United States, for example, the war effort led to the creation of the Council of National Defense. This government agency coordinated private industry, labor unions, farming, and transportation to increase the efficiency of wartime production. It also included an Educational Propaganda Department charged with managing "patriotic education."[56] Though coordination of this type was not entirely unknown in the past, its scope and intensity reflected the realization among US political leaders that the achievement of their strategic aims—the total defeat of Germany—required operations far beyond the battlefield. Their acknowledgment of this expanded battlespace generated insights and experience that would shape how the next war would be fought. But strategic awareness adaptation to a new strategic environment still would be required—a key point of this book as this historical review provides the reader with a lens through which to interpret the present critical juncture at which the United States stands in this regard.

World War II

It was widely understood by the mid-1930s that a second world war would incorporate many of the elements of the World War I battlespace, such as government coordination of national economies, strategic targeting of civilian populations, management of information on a societal scale, new ways of organizing military operations, and accelerated technical innovations in battlefield weaponry. Though the British science fiction writer H. G. Wells was off the mark in many of his predictions, his 1933 novel *The Shape of Things to Come* included a coming war in which a rearmed Germany and its adversaries bombed each other's cities, and submarines served as launching pads for "air missiles" to destroy faraway targets with weapons of mass destruction.[57]

In the more practical realm, from the mid-1920s through the 1930s, the American Joint Army and Navy Board considered the nature of military operations in the context of the changing geopolitical environment. It took these elements of the GRINS framework to develop War Plan Orange for a possible war with Japan in the Pacific. Envisioning the Japanese seizure of the Philippines (then a US Commonwealth), the plan foresaw the actual "island-hopping" campaign to recapture the Philippines.[58] The plan also foresaw the integration of air, land, and sea operations—the beginning of joint operations as an operational concept—and the necessity of supporting these operations with an integrated command structure. The Joint Chiefs of Staff were effectively set up in 1942, and later this organizational idea drove most notably the creation of the unified combatant commands (formalized in 1986) and specialized organizations such as the US Special Operations Command (1987).

War Plan Orange prioritized the role of the surface fleet with the aim of defeating the adversary's navy in a climatic sea battle. The influential ideas of Alfred Thayer Mahan, a former president of the US Naval War College, particularly in his *Influence of Sea Power upon History, 1660–1783*, shaped the conceptual realm in which War Plan Orange was devised.[59] The plan thus pursued US naval dominance to defeat Japan's battle fleet and to strangle Japan's economy. It did not foresee the operational importance of strategic air power—the Douhet-inspired bombing of Japan to destroy its industry in tandem with Mahan's idea of naval power to destroy the adversary's fleet. This interlocking of air, sea, and land (i.e., amphibious landings—the critical role of Marines) made the US advance across the Pacific possible, and strategic bombing in the form of the nuclear attacks on Hiroshima and Nagasaki marked the decisive end of the war in the Pacific.

The war in Europe presented different challenges. Technology and the organization of military operations (GRINS elements of innovations in science) were part of a shared battlespace. But Europe's geopolitical environment was different. At the outset (from a US perspective, as by the time the United States formally entered the war in December

1941, Britain and Germany had been at war for over two years), US, British, Soviet, and other Allied forces debated operational approaches to defeating Nazi Germany. British conceptions of the battlespace, influenced by the still recent experience of horrible battlefield casualties in World War I, privileged a more indirect enveloping strategy focused on the peripheries of German-controlled territory versus an American and Soviet preference for concentrations of forces to hasten the collapse of German forces.

This difference in operational approaches has important GRINS lessons for thinking about the contemporary critical juncture for US strategy. British strategists feared a repetition of World War I's static warfare and planned on the basis of their own relative weakness. This political concern, along with their more limited industrial capacity relative to their American allies and German adversaries, pushed them to prioritize indirect means. These means included support for partisans and uprisings in occupied territories that shaped the character of warfare in that territorial and conceptual arena of World War II. At the same time, the invasion of the European continent at Normandy (Operation Overlord; D-Day) was the central element of American strategy. In contrast to British plans (though British forces participated in the invasion), the direct attack intended to engage the German army was dispersed across the front as a consequence of Allied deception as to where the attack would occur so that a concentrated mass of military force could be used to smash the German forces. This military operation also highlighted the critical role of tactical air power to gain air superiority, isolate the battlefield, and support ground forces, combined with the application of strategic air power to destroy Germany's capacity and will to fight.[60]

Unconditional German surrender reflected American and Soviet, as well as British, approaches to victory. But the European war also prompted debate over the appropriate target of strategic bombardment and whether to prioritize indiscriminate (Douhet doctrine) or discriminate bombing of German infrastructure, industry, and people. Britain's Royal Air Force planners preferred indiscriminate bombing of German urban areas to break popular morale. US forces favored more discriminate targeting of economic infrastructure (electricity, petroleum, transportation) to destroy Germany's economy. This required the development of what were still rudimentary daylight precision-bombing technologies and operational techniques. This approach to strategic bombing was premised on the idea that fragile modern systems are vulnerable to air power.

Though there was considerable debate about the impact of strategic bombing, US air power ultimately played a decisive role in destroying petroleum and chemicals that were important for ammunition production. This use of air power validated the idea that General Mitchell advanced in the 1920s when he thought that American politicians and military operations planners did not grasp the importance of the air domain and how it was changing the battlespace. Forceful in his views, he was court-martialed for insubordination.[61] His fate illustrates the difficulties that strategic and operational innovators can face when organizational and political structures remain tenaciously committed to established doctrine and procedures. This is a lesson for our own time.

Conclusion

This application of the GRINS framework to the character of warfare in the nineteenth and twentieth centuries draws attention to elements of the broader political, technological, and organizational context that are important for thinking about the contemporary character of warfare. As the reader will discover later in this book, this framework will be applied to interpret the true character of contemporary warfare and how current policies and approaches to warfare are blind to key elements of this reality.

The first lesson is that the scope of warfare—the battlespace—is reflected in components of the GRINS framework that are salient in a particular time and context—that is, the character of warfare. Attention to regime type and ideas draws attention to the role in World War I of American "educational propaganda" to maintain political support for war fighting. Public opinion always has been important in warfare but not in this bureaucratically managed relationship to a political regime. Operational and technical innovations changed and intensified the place of civilians in the battlespace as concepts like strategic bombing became feasible.

A second lesson is that historical experience often shapes how planners and operational experts apply elements of the GRINS framework when they interpret and respond to the character of warfare in their time. This is a call to explore more deeply and systematically the old adage that countries often prepare to fight their most recent war when confronted with the next. British politicians and military commanders confronting Nazi Germany thought a great deal about the horrible toll that static warfare twenty-five years earlier had taken on their population and how it shaped public opinion. Their focus on an enveloping strategy and other indirect approaches reflected this legacy. Likewise, Mahan's influence on American strategic thinking, particularly on the central role of power in the maritime domain over the past three centuries, partially distracted the Americans from emerging advantages of combined joint operations and the strategic aspects of air power. These GRINS lessons will be applied later in this book when we explore the character of contemporary warfare.

Notes

1. Rosen, *Winning the Next War*.
2. Wahl and Toppel, *The Gatling Gun*, 11.
3. Headrick, *The Tools of Empire*, 118.
4. Clausewitz, *On War*, 75.
5. There is a scholarly enterprise dedicated to debating the changing nature and character of war. For example, Duyvesteyn and Angstrom, *Rethinking the Nature of War*; Ben-Ari, *Rethinking Contemporary Warfare*; Strachan and Scheipers, *The Changing Character of War*; Newman, *Understanding Civil Wars*.
6. Christopher Mewett, "Understanding War's Enduring Nature Alongside Its Changing Character," *War on the Rocks*, January 21, 2014.
7. Hoffman, "Squaring Clausewitz's Trinity in the Age of Autonomous Weapons."
8. Jomini, *Précis de l'art de la guerre*.
9. Berlin, *Against the Current*, 74.
10. Reddaway, *Frederick the Great and the Rise of Prussia*, 282.
11. Strachan, *Carl von Clausewitz's On War*.
12. Clausewitz, *On War*, 89.
13. Mann, *The Sources of Social Power*, 4 vols.
14. Pitt, "Warfare and Hominid Brain Evolution."
15. Dating back to 8,000–10,000 BC, there are archaeological signs in East Africa of humans waging socially organized wars, likely for resources. Lahr, "Intergroup Violence Among Early Holocene Hunter-Gatherers of West Turkana, Kenya."
16. Millett, Murray, and Watman, "The Effectiveness of Military Organizations."
17. Millett and Murray, *Military Effectiveness*, 3 vols.
18. Rosen, "Military Effectiveness."
19. Brooks and Stanley, *Creating Military Power*.
20. Biddle, *Military Power*.
21. Howard, "The Forgotten Dimensions of Strategy."
22. Mann, *The Sources of Social Power*, vol. 4.
23. For example, Fearon, "Rationalist Explanations for War"; Gartzke, "War Is in the Error Term"; Berdal and Malone, *Greed and Grievance*.
24. Wilson, *The Thirty Years War*.
25. Jackson, *Quasi-states*, 32–49.
26. Mann, *The Sources of Social Power*, 1:25–26.
27. Carr, *The Twenty Years' Crisis, 1919–1939*.
28. Mann, *The Sources of Social Power*, vol. 2.

29. Spruyt, *The Sovereign State and Its Competitors*.
30. Barkey, *Bandits and Bureaucrats*.
31. For a detailed discussion, see Mann's *The Sources of Social Power*, an epic four-volume history of power, from prehistory to the present.
32. Mann, *The Sources of Social Power*, 1:22.
33. Acemoglu and Robinson, *Why Nations Fail*.
34. The Declaration of Independence, US Congress, July 4, 1776. While it is generally held up as a perfect guiding document, we also know that much of what was agreed to in that time only applied to wealthy white males.
35. Isaiah Berlin essay in Bailey, *The Broadview Anthology of Social and Political Thought*, 2:345.
36. Berlin, *Two Concepts of Liberty*, 119.
37. Berlin, *Two Concepts of Liberty*, 119.
38. Saideman and Zahar, *Intra-state Conflict, Governments and Security*.
39. Janowitz, *The Professional Soldier*.
40. Mansoor and Murray, *The Culture of Military Organizations*.
41. Lasswell, "The Garrison State."
42. Machiavelli, *The Prince*, 77.
43. Friedberg, "Why Didn't the United States Become a Garrison State?"
44. Donnithorne, *Four Guardians*.
45. Vincent of Lérins, *Commonitorium*, trans. Reginald S. Moxon, chap. 2, sect. 6, Nicene and Post-Nicene Fathers, Series II, Vol. XI [430 AD], 132.
46. Andress, *The Terror*.
47. Mikaberidze, *The Napoleonic Wars*.
48. Weber, *Peasants into Frenchmen*.
49. Charles Tilly, "Reflections on the History of European State Making," in Tilly, *The Formation of National States in Western Europe*, 42; see also Ikenberry, *The State*, 40–41.
50. Klar, *The French Revolution, Napoleon, and the Republic*.
51. Jünger, *Storm of Steel*.
52. Van Evera, "The Cult of the Offensive and the Origins of the First World War."
53. Matisek, Robison, and Jayamaha, "Extending the American Century."
54. Douhet, *The Command of the Air*.
55. Liddell-Hart, *A History of the World War, 1914–1918*.
56. Record Group 62, Records of the Council of National Defense, National Archives, www.archives.gov/research/guide-fed-records/groups/062.html.
57. H. G. Wells, "Changes in War Practice After the World War," in *Shape of Things to Come* (London: Hutchinson, 1933). Available at Project Gutenberg, http://gutenberg.net.au/ebooks03/0301391h.html#chap2_04.
58. Miller, *War Plan Orange*.
59. Mahan, *The Influence of Sea Power upon History*.
60. A particularly useful explanation of this concept is found at War Department, "Tactical and Strategic Air Power in World War II 'Air Power and Armies' Reel 1 76704," video uploaded to YouTube by PeriscopeFilm, September 14, 2017, www.youtube.com/watch?v=oXNXfWqn7M0.
61. "American General William D Mitchell Speaks on Importance of Air Power," 1936, video uploaded to YouTube by Critical Past, June 14, 2014, www.youtube.com/watch?v=wuWLh4mkePE.

3
Warfare in the Nuclear Age

AT THE END OF WORLD WAR II, THE INTRODUCTION OF NUCLEAR weapons gave major powers an awesome capability for destructive power that was reflected in popular culture through images like the cataclysmic end of civilization in the 1964 Stanley Kubrick film *Dr. Strangelove or: How I Learned to Stop Worrying and Love the Bomb*. The destructiveness of this weapon raised an existential question: How can such a weapon be used in combat to achieve strategic aims when the result is likely to be the annihilation of both the user and the adversary?

Initially some conceived of nuclear weapons as simply more powerful versions of conventional bombs, with similar operational and strategic effects. Gen. Curtis LeMay, head of the Strategic Air Command from 1948 to 1957, considered using nuclear bombs on North Korean cities during the Korean War to deliver a decisive blow to end the conflict. He questioned whether there would have been more decisive strategic effects with nuclear weapons in light of the huge civilian casualties that actual US operations inflicted on North Korea through the use of napalm and high-explosive weapons in urban areas and precision bombing of dams and other infrastructure vital for agricultural production.

From the GRINS perspective the technical innovation of nuclear weapons prompted military operational innovation in the form of a centralized command-and-control system to maintain an unprecedented continuous readiness capability in peacetime and the search for new ways to base nuclear weapons on land and at sea.

But nuclear weapons were part of a strategic revolution, an ideational shift in GRINS terms. Their power in the context of Cold War geopolitical competition offered the prospect of destroying an enemy beyond any rational purpose and the likelihood that nuclear retaliation before that point would destroy the United States as well. Bernard Brodie noted in his 1959 classic *Strategy in the Missile Age*, "We should also recognize once and for all that when it comes to predicting human casualties, we are talking about a catastrophe for which it is impossible to set upper limits appreciably short of the entire population of a nation."[1]

The unexpectedly high fifteen-megaton yield in the 1954 Castle Bravo test of a thermonuclear device, a thousand times more powerful than the nuclear weapons dropped on Hiroshima and Nagasaki in 1945, reinforced the notion that this weapon meant that force had to be exercised in different ways. Devices of destructive power on a scale large enough to wipe out large cities and spread radioactive fallout over vast areas would overwhelm conventional strategic-bombing distinctions between industrial, infrastructural, and populated targets. The Castle Bravo test also pushed the dangers of nuclear fallout firmly into the public mind, adding a new ideational dimension to nuclear warfare. Strategy now had to achieve the prevention of nuclear conflict as well as the use of force in military operations. This revolution was encapsulated in the strategy of deterrence, a foundational element of US Cold War policy to contain Soviet expansion. This shifted the US definition of strategy from the use of combat to pursue victory to the use of military force to deter war.

National Security Council Paper 68, titled "United States Objectives and Programs for National Security," defined this stalemate in 1950 as containment of the Soviet Union's power and confirmed, in keeping with earlier US official moves, that avoidance of the use of its most potent weapon was the ultimate strategic priority.[2] The practical issue then became how to define and structure a deterrent-containment force and how to pay for it.

Warfare and the Strategic Paradox of the Cold War

The strategic paradox that faced the United States and the Soviet Union was that both defined their ultimate strategic aims in terms of forcing their adversary to bend to their political will. The nature of warfare remained constant in this regard. But victory prior to the nuclear age context would translate into something like what the United States and the Soviet Union enjoyed in their respective spheres of influence after their joint decisive defeat of Nazi Germany. Now the ideationally and physically awesome power of nuclear weapons weakened the link between political and military objectives. In short, these countries now possessed hitherto unimaginable capacities to use force, but the utility of this force was in doubt.

This strategic paradox was the foundation of a new geopolitical framework for competition, expressed in terms of a bipolar world, the maintenance of deterrence against the use of nuclear weapons, and the indirect use of force in the pursuit of strategic aims.

Preventing the use of nuclear weapons defined strategic debates. Most Americans at the end of World War II viewed war as a realistic option in the service of national interests and military operations as a science of victory. But the strategic paradox of the world of nuclear weapons meant that military force was not so much exercised as threatened. Thomas Schelling explained that nuclear weapons could be utilized in a diplomacy of violence—the credible threat to use this awesome force—to preserve peace. This meant that military capabilities, whether real or imagined in the minds of the adversary, were to be used as bargaining power. Preparations for armed conflict, such as steps taken by the United States during the Berlin and Cuban crises in the early 1960s, were not merely preparations for military engagements. These were signals to an adversary. In this battlespace the reports of the adversary's own military intelligence were as important as the United States' most important diplomatic communications.[3] Armed deterrence of this sort required skillful diplomacy and careful control and coordination of military operations. This shift in the GRINS approach to military operations was necessary, as even tactical-level operations could have strategic consequences if they were interpreted by the adversary as signals.

The demarcation of areas of vital interest underpinned this stalemate. The establishment of the North Atlantic Treaty Organization (NATO) in 1949 facilitated the creation of other intra-European multilateral institutions. The American sphere of influence, referred to as the "free world," was shaped not by ideational purity but by pragmatism. It included Portugal, Spain, and Greece, dictatorships in which citizens were not particularly "free," and France and Britain, which at the founding of NATO still held extensive colonial empires. This strategic alliance was buttressed and defended by a direct and explicit American security umbrella.[4] The Soviet Union established its sphere of direct interest and influence with its own security umbrella, which encompassed most of eastern Europe. From this came the Warsaw Pact, designed to counter NATO, as the American-led rearmament of

West Germany brought back geopolitical memories of the dangers of a remilitarized Germany.[5]

The actual use of armed force, as opposed to the threat of force, during the Cold War rested on a strategic doctrine of limited war.[6] This was an indirect means to manage the strategic paradox on the periphery of areas of vital interest. Limited war was not a new concept, but its elevation to such a central place in doctrine was a novel development. This development reflected the strategic paradox in which direct confrontation was likely to end in mutual destruction. The alternative was to pursue strategic objectives through indirect means. In GRINS terms, this radical technological and ideational shift made strategically relevant the many local wars and insurgencies around the world after World War II.

Recall that Britain defined support for partisan resistance and involvement in other local wars in the context of the wider World War II (as did the United States in a more limited fashion) as an important element of its strategy of envelopment. The emergence of and reliance on proxies for indirect warfare was not a new phenomenon; it was a time-honored tradition in conflicts dating at least back to the Peloponnesian War.[7] But the technological, ideational, and geopolitical GRINS realities of the nuclear age meant that the great majority of these now strategically relevant conflicts did not directly involve a nuclear or even a major power. Nor were any of these conflicts fought on the territory of a major power.

This consistent and prolonged territorial displacement of limited warfare was accompanied by a distinctive US and Soviet willingness to deliberately refrain from conducting military operations in support of their proxies anywhere near their full capabilities. Great care was taken to ensure that US and Soviet military personnel did not directly confront each other to guard against the possibility of escalation that could lead to nuclear war. In the rare instances when the two adversaries' military forces engaged, Schelling's "diplomacy of violence" was deployed to mute any unintentional signal of escalation. In one instance during the Korean War, Soviet and US pilots engaged in air-to-air combat. Both sides kept it private, an example of denial of actual military force in the interstitial space at the tactical level. Even an accidental US attack on an airbase in the Soviet Union during the Korean War led to diplomatic efforts on both sides to reduce the likelihood of this provocation spiraling out of control.[8]

Equally important was the discipline exercised in not initiating any wars in places that the adversary defined as an area of vital interest. When uprisings or insurgencies did appear in these areas, such as the

1956 Hungarian Revolution, the opposing nuclear power took no militarily meaningful action. Soviet use of force in Czechoslovakia to crush protests and proposed political reforms in that country solidified what became known as the Brezhnev Doctrine, which held that the Soviet Union had the right to intervene in any country where a Communist government had been threatened. Tacit mutual acceptance of this doctrine did not rule out very low-level material and somewhat higher-volume rhetorical support for insurgents in these places.

The United States and the Soviet Union accepted a rough stalemate to avoid annihilation at the same time that they sought to use force to further their strategic interests. Exercise of violence in this strategic milieu gave rise to a form of warfare of a very different character, best conceived as "interstitial warfare," which took place in the parts of the world most generally referred to at the time as the "Third World," which neither nuclear power defined through its actions as an area of vital interest. This region is where the use of force, rather than simply the threat of force, had practical utility.

Interstitial Warfare: A Nuclear Age Indirect Approach

Limited warfare on the part of nuclear-armed states in pursuit of political aims against another nuclear-armed state departed from Carl von Clausewitz's original formulation that the military means of a conflict are proportional to its political end.[9] Clausewitz's expectation of proportionality of ends and (limited) means holds up outside this nuclear shadow, such as with American military interventions in small Central American and Caribbean states through much of the twentieth century. Changing the political scene in these countries did not require applying all available military force or even smashing the opposing force. Likewise, Russia's seizure of the lightly defended Ukrainian province of Crimea in 2014 did not require that Russia destroy Ukraine's armed forces to seize that territory and population from a sovereign state in pursuit of its ultimate political objective.

Cold War practice among nuclear-armed states diverged from Clausewitz's formulation because, as noted above, limited warfare was essential to avoid the introduction of nuclear weapons into a conflict. Yet the political ends of these conflicts ultimately concerned the struggle between the United States and the Soviet Union. The improbability of using force in areas of each country's vital interests made virtually any military conflict in the (generally "Third World") interstitial zone reflective of the political confrontation between the United States and the Soviet Union. This condition left the two adversaries with a much more complex and unstable linkage between their political objectives in engaging in these conflicts and their military means.

In practical terms, limited warfare meant seeking political gains indirectly through the application of force against the adversary's client states outside the zones of vital interests. When either of the two adversaries actively engaged in such a conflict, they found themselves allied with politically unstable governments. In South Korea and South Vietnam, the United States supported deeply corrupt and unpopular governments that faced opposing armies and guerrilla insurgencies seeking the reunification of their countries. Soviet forces in Afghanistan in the 1980s faced a similar situation. Despite this local (and often deeply parochial) aspect of these conflicts, they were viewed in US and Soviet circles as extensions of their political struggle with one another. "If we let the Koreans down," President Harry S. Truman told Congress after the June 1950 North Korean invasion of the South, "the Soviets will keep right on going and swallow up one piece of Asia after another. . . . If we let Asia go, the Near East would collapse, and there is no telling what would happen in Europe."[10]

Truman's evocation of the "Domino theory" (that countries will fall to communism like dominos) spoke to core national interests, but in terms of this or any other interstitial war, it ultimately left the intervening force with ambiguous political aims because there could be no clear victory without risking nuclear apocalypse. In a pattern that would appear in American wars after the Cold War—but for different reasons—the immediate political ends of engagement in interstitial wars thus were prone to revision. Stripped of the capacity to apply full military capacities when campaigns did not proceed as planned—a core element of the nature of warfare—political rationales shifted, usually in a downward trajectory. Was the political end to unify the country? Was it to hold the line? Was it to leave the ally to its own devices with US support for its military capacity?

It is not surprising that this ambiguity in the means and ends of these interstitial wars encountered popular dissatisfaction, particularly when the cumulative costs, both in casualties and budgetary outlays, steadily mounted. The United States' political system accommodated questions about the perceived political value of these engagements of US military force. But these popular sentiments were present in the Soviet Union too. The popular 2005 Russian film *9 Poma* (9th company), set during the Soviet-Afghan war in the 1980s, captures the frustrations of soldiers deployed to fight for ambiguous, shifting, and ultimately illusory political goals.[11] The defection of KGB archivist Vasili

Mitrokhin with a trove of documents that brought to light internal Soviet discussions of these problems underscored the extent to which both adversaries grappled with these difficulties of interstitial warfare.[12]

These tensions between ambiguous political ends and limited military means influenced the development of US doctrine. The Powell Doctrine called for the restoration of defined ends and the use of overwhelming means to achieve victory with a minimum of casualties. Its first version appeared in the 1992 national military strategy and reflected purported lessons from the 1991 Gulf War. But it was understood in the context of Vietnam:

> Once a decision for military action has been made, half measures and confused objectives exact a severe price in the form of a protracted conflict which can cause needless waste of human lives and material resources, a divided nation at home, and defeat. Therefore one of the essential elements of our national military strategy is the ability to rapidly assemble the forces needed to win—the concept of applying decisive force to overwhelm our adversaries and thereby terminate conflicts swiftly with minimum loss of life.[13]

This doctrine of direct engagement was not feasible prior to the collapse of the Soviet Union, but its adoption in the early 1990s gives insight into the challenges that interstitial warfare posed in domestic American politics and in civil-military relations. But remaining within the Cold War context, the adversaries compensated for these difficulties through other approaches to the interstitial condition and the larger strategic paradox of nuclear weapons. These measures included indirect and often deniable proxy support to wear down and preoccupy the strategic adversary.

Indirect Warfare:
Attrition in the Geostrategic Periphery

The geostrategic element of the GRINS framework—the division of the world into two static spheres of vital interest to the nuclear-armed United States and Soviet Union—pushed armed conflict to the confines of the strategic periphery. Unlike in wars in Korea, Vietnam, and Afghanistan, the Cold War adversaries avoided the application of their own military forces to these conflicts. That decision considerably reduced chances of nuclear escalation. As B. H. Liddell Hart wrote, "The nuclear deterrent does not apply and cannot be applied to the deterrence of subtler forms of aggression."[14] It also addressed the problem of popular concern about casualties and costs of fighting in wars for ambiguous ends.

This mutual accommodation on both sides of the nuclear divide to the strategic paradox of the Cold War translated into the conduct of warfare through proxies of limited means. This involved the provision of weapons, financial assistance, and political support to third parties, whether armed forces of states or nonstate actors ("rebels"), that did the actual fighting while absolving the sponsor of the direct introduction of force. The desired political effects included the protection and extension of influence in peripheral regions through the medium of localized wars. The overall political effect could include the perception that the rival political ideology was blocked in a particular global region and the progressive exhaustion of the adversary.

The use of proxy warfare to influence strategic balances was a familiar practice long before the Cold War. Indeed, indirect influence through support for surrogates was a foundation of the British capacity to avoid the use of expeditionary forces in warfare on the European continent for three centuries before World War I. The Cold War–era use of proxies was consistent with this underlying strategic nature of warfare, focused as it was on achieving desired political results of warfare through this indirect means. Surrogates also consistently pursued their own agendas then, as they do now, including the recruitment of powerful patrons that they attempt to leverage for local advantage. Surrogates can (and often do) deceive and exploit asymmetries of information about their own societies and political relationships to entangle patrons in their affairs and extract resources on favorable terms.

The character of proxy warfare during the Cold War, however, was distinct. Shared US and Soviet interests in limiting warfare to the geostrategic periphery led to stalemates in means and ends across multiple wars in the strategic periphery. This outcome reflected what Stathis Kalyvas and Laia Balcells called the "technology" of Cold War proxy warfare in which external patrons would raise the capacities of their surrogates, whether states or rebels, to balance the adversary's surrogates. This improvement of surrogates' capacities would include material support as well as subtler means, such as ideological training and networking, support in international institutions, and doctrinal advice. Rebels tended to benefit more than states in these conflicts, as they usually needed more help from their patrons to remain in the fight.[15]

This balancing dynamic had the effect of lengthening civil wars as stalemates ensued. Wars started at a steady rate between 1945 and 1989, but as their duration lengthened, the number of ongoing wars grew. According to one dataset, Cold War conflicts peaked at forty-seven in 1987, compared to fewer than fifteen through much of the 1950s.[16]

Notable examples of prolonged civil wars included Angola (1961–1999) and Mozambique (1961–1992), which began as indigenous struggles against colonial rule. By the mid-1970s, they had evolved into intense confrontations between Soviet-backed states and US-backed rebels. This situation created a general impression among observers that it was very difficult for state forces to defeat rebels, despite the substantial record of successful counterinsurgencies prior to the Cold War.

Notably missing from the conduct of indirect warfare in the periphery on the American side was the idea that support for a proxy was instrumental and that the relationship did not necessarily require a government or rebel group to reflect American political ideals in its internal function. Rhetoric mattered, especially in international forums, but American support involved considerations of regime change only insofar as it would replace a less effective client with a more effective (usually less corrupt or factionalized) alternative. The US backed coup in 1963 against South Vietnamese president Ngô Đình Diệm was one such example.[17] The ousted (and promptly executed) president's successor was hardly more democratic but was more compliant with US interests. The US role in this affair bore little hint of the post–Cold War American habit of forcible regime change as part of an agenda to replace authoritarian rule with institutions of democratic governance.

Though Soviet backing also reflected pragmatic calculations, Soviet officials often extracted a heavy price in return for their support by requiring beneficiaries to emulate Soviet institutions and policies that often were spectacularly unsuited to the economic conditions of very poor countries and the political preferences of their populations. American strategists exploited this difference in approaches. As Dean Acheson, US secretary of state from 1949 to 1953 and key designer of early–Cold War US foreign policy, pointed out, visible attempts to alter the government of an ally or partner would damage US credibility while undermining the legitimacy and authority of that government in the eyes of its citizenry.[18] This approach equipped the United States to present itself as a defender of the "free world," which included many dictatorships, against a rigid totalitarian (and, over time, economically much less dynamic) alternative.

Political Warfare

George Kennan, the State Department's first policy-planning chief, responded to this dilemma with proposals to pursue political warfare, "the employment of all the means at a nation's command, short of war, to achieve its national objectives."[19] These means ranged from political alliances, economic measures, and propaganda to covert and clandestine means of psychological warfare and subversion. Kennan observed that the United States needed to take more seriously covert political warfare under the direction of the Department of State rather than government departments commonly associated with waging war. He proposed (1) supporting liberation committees of refugees who had fled Communist nations, (2) underground activities behind the Iron Curtain with "remote and deeply concealed official control of clandestine operations so that government responsibility cannot be shown," and (3) support for indigenous anti-Communist elements in threatened countries of the free world.

Kennan advocated the use of private intermediaries as surrogates for US government agencies to support anti-Communist organizations. He called for the creation of a political warfare operations directorate within the government outside the Department of State because the operations of the directorate needed to be concealed in a covert budget.

Specific attention to subtler and more diverse ways to achieve political ends included the use of new international organizations such as the United Nations and regional bodies such as the Organization of African Unity to influence global opinion and levy sanctions against the adversary's surrogate. This use of international law as a weapon of war was demonstrated with the decision of the United Nations or other international bodies to formally recognize a rebel group as the "sole legitimate representative of the people" of a state in question, for example, strengthening the political influence of the group's patron. Political recognition of the proxy's claims was about calling upon other governments to take sides to strengthen a beneficiary and isolate the opposing domestic force in ways that disadvantaged the nuclear-armed adversary's strategic position in the periphery. Alarmed US State Department officials reported, for example, that the Soviet-backed "guerrilla group" fighting for the liberation of Guinea-Bissau from Portuguese colonial rule had declared unilateral independence in September 1973 and immediately was recognized by "38 nations, including the Soviet Union, the [People's Republic of China], Romania, Yugoslavia and a score of African states," a number expected to grow.[20]

US officials recognized the role that federal coordination of information warfare played in engaging in a global struggle for hearts and minds. Established in 1953 (and disbanded in 1999), the United States Information Agency (USIA) aimed to use "public diplomacy" to "combat weapons of false propaganda and misinformation" through the coordination of positive information about the United States through cultural and educational exchanges and broadcasters such as Voice of America to spark the desires of individuals in Communist-dominated societies for democratic values and free access to information.[21]

This Cold War expansion of the scope of competition into the ideational space included the realm of higher education, particularly insofar as scholars produced knowledge about Third World societies. The 1958 National Defense Education Act created a federal program to fund area studies programs in the United States. It initially funded nineteen centers focused on area and international studies under the framework of Title VI National Resource Centers to offer less commonly taught languages and support cultural exchanges with "strategic world regions."[22] The Fulbright-Hays Act of 1961 established a higher-education exchange program that initially was administered by USIA to "demonstrate the possibilities and values of cooperative action . . . and pave the way for closer and more fruitful political relations."[23] Nongovernment influence operations played a role too. The Ford Foundation, for example, supported educational and cultural exchanges through the 1950s that addressed fears "that Marxism and Communism would exert a growing appeal among disaffected intellectuals" in war-ravaged Europe and "emerging nations" of the Third World.[24]

Soviet political warfare operations covered a similar spectrum, though the ways and means of operations reflected the centralized and authoritarian nature of the Soviet Union's regime. The Soviet Union maintained a large and complex bureaucracy for these purposes that included the KGB, the International Department of the Communist Party of the Soviet Union, the International Information Department of the Communist Party, and a plethora of actors such as journalists, scholars, and students whose links to the Soviet state were not always apparent. The overall objective of these measures was to exploit the relative openness of Western societies and diversity of information outlets to deceive and confuse individuals and governments and cultivate a more positive image for Soviet policies.

Many of these Soviet operations focused on what were called "active measures" to weaken the West by creating doubts in people's minds about the true nature of the United States and other Western governments. Examples of these operations included disinformation campaigns, such as planting rumors in western European academic and government circles that the Central Intelligence Agency was planning a coup in their country, that the United States was behind an Islamist "false flag" attack on the

Grand Mosque in Saudi Arabia in 1979, and that the United States was behind the death of Panamanian leader Omar Torrijos in 1981. Forgeries, such as seemingly official US war-fighting plans, were distributed to create distrust of US official statements. Fake communiqués aided efforts to manipulate press reports to portray the US role in the Camp David process in the early 1980s in a negative light. Soviet clandestine support for solidarity committees, such as European-based support groups for pro-Soviet insurgencies in Third World countries, helped to call into question and undermine official policies in democratic societies with strategic effects. Political influence operations targeted journalists to push pro-Soviet positions on a range of issues in a subtle fashion. In other cases, Soviet officials also established close personal relations with foreign officials, for instance, playing on their self-importance with invitations to meet Soviet leaders.[25]

On occasion Soviet political influence operations had significant political effects within the US zone of vital interests. Soviet support for West Germany's *Ostpolitik* efforts to come to a rapprochement with East Germany and its Soviet backers gave Soviet officials an opportunity to exploit differences between West Germany's definition of strategic interests as a core member of NATO and emerging strategic interests that reflected its geographical proximity to the Soviet sphere in Europe. West German chancellor Willy Brandt resigned in 1974 after his secretary, Günter Guillaume, was discovered to have been an agent of the East German Stasi (secret police) from the 1950s and to have regularly passed documents from the Chancellor's Office to his East German handlers.[26] Declassified documents show the role of Soviet subversion in connection with demonstrations in West Germany against the stationing of medium-range nuclear missiles in the 1970s and 1980s that plagued this significant NATO effort to address the nuclear strategic balance in Europe.[27]

Soviet officials attempted to use the ideational space of higher education in ways that were not entirely different from US efforts. The People's Friendship University, founded in 1960 and rechristened the Lumumba Friendship University in 1961, was designed to build international networks of personal relationships and propagate ideas and impressions that put the Soviet system and its role in the world in a positive light. Though alumnae networks were considerably less extensive than those of US universities and exchange programs like Fulbright-Hays, Lumumba Friendship University alumnae included influential people such as Nicaraguan president Daniel Ortega, Iranian supreme leader Ali Khamenei, Palestinian National Authority president Mahmoud Abbas, and Venezuelan terrorist Illich Ramírez Sanchez ("Carlos the Jackal").

Though Soviet universities and exchange programs usually were less influential than their American counterparts as instruments of political warfare, Soviet planners and operatives had an advantage in efforts targeting open societies. This is another instance in which the regime element of the GRINS framework plays a significant role in shaping the character of warfare. There was no Soviet counterpart to western European tolerance for Soviet-aligned Communist parties. Italy's Soviet-aligned Communist Party regularly captured about 30 percent of the vote in elections before its break with Moscow in the late 1960s. That party was the target of clandestine American and British efforts to influence the Italian electorate, an example of political warfare within a nuclear-armed country's own zone of vital interests.

Ideationally, the United States and its allies were at a disadvantage. The Soviet Union and its satellite states in Eastern Europe had built ideational, economic, and political defenses to inoculate citizens living inside their closed societies. Western states engaged in espionage activity, but conducting acts of political subversion inside Warsaw Pact countries was extremely difficult, if not impossible, given the respective police state each had, with little civil society as it was understood in democratic countries. The Communist state controlled all aspects of life, to include language, rhetoric, books, music, movies, art, and so forth—all to ensure a unity of Communist messaging and indoctrination. At the same time, Western universities and other influencers of public discourse included Marxists who openly engaged in antiliberalist dialogue, precisely because a free society encourages open discourse. Soviet agents took advantage of this discourse to highlight very real social and political problems, such as systematic racial discrimination in the United States.[28]

Cold War Military Organizational Reform and Innovations

As the United States carried the lion's share of the security burden in both Europe and the Pacific, its armed forces relied on the legacy organizational structure inherited from World War II. Organizational reformers focused on how to fight the Cold War better and, in the process, created the institutional structure that shapes the US military in the 2020s. As the 1980s dawned, the United States turned its conscript citizen army into an all-volunteer force. In spite of adapting advances in technology to existing military organizational structures, the US military faced a series of failures (e.g., Vietnam, Operation Eagle Claw in Iran) and near setbacks (Operation Urgent Fury in Grenada). The lessons learned led to a fundamental reorganization of the US armed forces in the form of the 1986 Goldwater-Nichols Act, which has had far-reaching ramifications for how the United States has fought wars since that reform.

In line with the nature of the Cold War global strategic contestation, the US military was reorganized along geographic unified combatant commands and unified functional commands that transcend geographic boundaries. It also removed the service chiefs from the operational chain of command to an administrative chain of command, instilling a clear division of labor between combatant commands and services. In turn, the role of the highest-ranking military officer, the chairman of the Joint Chiefs of Staff (CJCS), was transformed into an advisory one. Consequently, service chiefs, serving as the highest-ranking members of their respective service branches, were tasked with the explicit mission of organizing, training, and equipping their personnel to be combat ready. Combatant commands—the Joint Force Commanders—would optimally use the proper blend of personnel from each branch to perform the actual war-fighting mission. The operational chain of command runs directly from the combatant commander to the secretary of defense to the president, with service chiefs and the CJCS playing only an advisory role.

Just as US forces gained an absolute quantitative and qualitative edge vis-à-vis the Soviets, the Cold War ended in the form of an internal implosion of the Soviet Empire. This was followed by a dismantling of its alliance structure in all its spheres of direct interests, influence, and contestation. The animating ideology of the Soviet Union had lost its traction, and its economy was stagnating. Its authoritarian political system in an era of satellite TV could no longer hide the affluence and progress in adversarial states from its own citizens. It was not so much that the Soviet Union was competing against the European or the US economy by the 1980s; it was already falling behind newly prosperous South Korea, Taiwan, and Singapore. The Soviets retained their military preponderance, to include about 37,000 nuclear weapons, but had lost their technological edge due to corruption and inefficient allocation of resources.

The Soviet system had become so rigid and ossified, its structures so brittle, that attempts at internal reorganization through glasnost (openness) and perestroika (reform) brought down the overall edifice.[29] It was as if all the internal ideational, economic, political, and military contradictions inherent in the Soviet system—as articulated by George Kennan—which overwhelming violence and terror had held at bay for so long, had finally coalesced, leading to an implosion.[30] Arthur Koestler presciently captured the human reality in 1939, at a time when the ideational power of communism was ascendant and would soon become the overarching ideology that would pervade most of the developed and developing world for a generation:

> The Party denied the free will of the individual—and at the same time it exacted his willing self-sacrifice. It denied his capacity to choose between two alternatives—and at the same time it demanded that he should constantly choose the right one. It denied his power to distinguish good and evil—and at the same time spoke pathetically of guilt and treachery. The individual stood under the sign of economic fatality, a wheel in a clockwork which had been wound up for all eternity and could not be stopped or influenced—and the Party demanded that the wheel should revolt against the clockwork and change its course. There was somewhere an error in the calculation; the equation did not work out.[31]

Conclusion

The GRINS framework helps to understand how changes in geopolitics, regimes, ideas, the nature of military organization, and scientific knowledge (technology) shaped the changing character of warfare. For the first time in history, a weapons system in the form of nuclear weapons played an endogenous role in shaping both strategic and geopolitical realities. An altered strategic and geopolitical reality refashioned the choices available to political and military leaders that sought to use violence as a means to generate military or political effects. In doing so, they backstopped conventional forces with a nuclear deterrence in the direct spheres of interest. That shifted the locus of confrontation to spheres of influence and contestation, where the superpowers, with their proxies, used the strategic incongruities inherent in the Cold War to engage in interstitial attrition, which became the defining character of warfare in this era.

Civil wars in seemingly distant lands—that is, places distant from Moscow and Washington, whose citizens would be hard put to find them on a map, such as Angola, Namibia, Oman, El Salvador, and Guatemala—with localized grievances and objectives were often tied into the broader bipolar strategic logic.

The shift in the type of warfare from direct confrontation to an interstitial form and the locus of warfare away from the direct spheres also indirectly had positive regional and global ramifications. Smaller countries did not have to worry about being absorbed by their larger neighbors through military force as had been the reality for centuries—perhaps since the beginning of recorded history. Most in the Western spheres used the external stability that comes from Western security guarantees to turn their islands of stability into islands of prosperity, while those allied with the Soviet Union tuned their stability into oppression, scarcity, and dread. The dénouement of the Cold War presented in the eyes of many a clear triumph of democracy and capitalism over Communist one-party dictatorships and socialist economic systems. This interpretation of the end became the dominant explanatory narrative. Remarkably, this interpretation omitted the role of overt and covert exercise of violence, the security architecture that became both the necessary condition and the security underwriter, and the strategic role all instruments of national power played in generating this desired end.

The Cold War also demonstrated the enduring nature of warfare. The essential connection between ways and means to the political ends of warfare remained firmly intact. Basic concepts such as indirect warfare endured. But the ideational and societal space in which nuclear-armed adversaries conducted warfare and its close strategic partner, deterrence, underwent considerable expansion. This expansion was particularly notable in the sphere of political warfare. But as the chapters that follow highlight, in the shadow of America's post–Cold War global hegemony, its political leadership, its foreign policy establishment, and many of its military planners largely forgot these lessons.

Notes

1. Brodie, *Strategy in the Missile Age*, 167.
2. "National Security Council Report, NSC 68, 'United States Objectives and Programs for National Security,'" Wilson Center, April 14, 1950, https://digitalarchive.wilsoncenter.org/document/116191.pdf?v=2699956db534c1821edefa61b8c13ffe.
3. Schelling, *Arms and Influence*.
4. Sayle, *Enduring Alliance*.
5. Mastny and Byrne, *A Cardboard Castle?*
6. Osgood, *Limited War*.
7. Proxy warfare was a way for Athens and the Peloponnese to compete against one another.
8. The US military would eventually punish the two pilots for their gross navigational error caused by stronger than normal winds aloft. For one pilot perspective, see Alton H. Quanbeck, "My Brief War with Russia," *Washington Post*, March 4, 1990.

9. Clausewitz, *On War*, book 1, chap. 2, "Purpose and Means of War," 90–98.
10. Quoted in Jablonsky, "The Persistence of Credibility," 8.
11. "9 Рота | Прилетели в Афган" (video uploaded to YouTube by GG TV, January 26, 2019, www.youtube.com/watch?v=Sx-CPwO3zm8) depicts a scene confronting a main character upon his arrival at Bagram. For a song expressing a similar sentiment, see "Афганские песни "Дойдем!" Э.Конт," video uploaded to YouTube by Svirochka, February 26, 2011, www.youtube.com/watch?v=1yFihWsB57k.
12. Andrew and Mitrokhin, *The World Was Going Our Way*. The Mitrokhin Archive can be accessed at Princeton University's Woodrow Wilson Center, https://digitalarchive.wilsoncenter.org/collection/52/mitrokhin-archive.
13. *National Military Strategy of the United States*, US Department of Defense, January 1992, https://history.defense.gov/Portals/70/Documents/nms/nms1992.pdf?ver=2014-06-25-123420-723.
14. Liddell Hart, *The Strategy of the Indirect Approach*, 373.
15. Kalyvas and Balcells, "International System and Technologies of Rebellion," 415–429.
16. Pettersson and Öberg, "Organized Violence, 1989–2019."
17. Cable 243, US Embassy–Saigon, National Security Archive, August 24, 1963, https://nsarchive2.gwu.edu//NSAEBB/NSAEBB101/vn02.pdf.
18. Acheson, *Present at the Creation*, 223–224, 356–257, 734–736.
19. Kennan, "Policy Planning Memorandum."
20. "Portuguese Guinea: Rebels Establish Government," Department of State, October 5, 1973.
21. Snyder, *Warriors of Disinformation*. For the text of Smith-Mundt Act of 1948, see https://govtrackus.s3.amazonaws.com/legislink/pdf/stat/62/STATUTE-62-Pg6.pdf.
22. Brecht and Rivers, *Language and National Security*.
23. Young, "Educational Exchanges and the National Interest," 17
24. McCarthy, "From Cold War to Cultural Development," 94.
25. "Soviet 'Active Measures': Forgery, Disinformation, Political Operations," Special Report no. 88 (Washington, DC: Department of State, October 1981), available at *Inside the Cold War*, http://insidethecoldwar.org/sites/default/files/documents/Soviet%20Active%20Measures%20Forgery,%20Disinformation,%20Political%20Operations%20October%201981.pdf.
26. Adams, *Strategic Intelligence in the Cold War and Beyond*, 58–62.
27. "Soviet Covert Action (the Forgery Offensive)," Hearings Before the Subcommittee on Oversight of the Permanent Select Committee on Intelligence, House of Representatives, Ninety-Sixth Congress, Second Session, February 6 and 19, 1980.
28. Woods, *Black Struggle, Red Scare*.
29. Meyer, "How the Threat (and the Coup) Collapsed."
30. Gaddis, *George F. Kennan*.
31. Koestler, *Darkness at Noon*, 257–258.

4

The Complexity of Emerging Battlespaces

HUMAN AGENCY IS A CONSTANT DEFINING FEATURE OF WARFARE since decisions of protagonists play a decisive role in shaping battlespaces. These decisions are contingent because not all protagonists—state and nonstate armed actors—get to pick all the choices they face. Some have the power and foresight to prepare for and shape their choices in a decisive manner, but even the powerful and wise face unexpected choices and make decisions that have unintended consequences. This was doubly so in the period of nearly comprehensive change following the end of the Cold War.

Europe and the United States emerged victorious at the end of the Cold War, which came to be referred to as the "unipolar moment," in which they exercised overwhelming military and ideational dominance. At that point, American and European actions played a decisive role in shaping the character of warfare. All of the features of the GRINS framework were in play: geopolitical realities, ideational currents, and American military hegemony, thus, by extension, regime types, defined the strategic environment. The preponderance of power concentrated in the West meant it could shape geopolitical realities both with the use of force and in the form of transnational juridical arrangements. These juridical arrangements reflected Western ideas, especially regarding diplomatic and economic power and the occasional use of military prowess for enforcement. International law, backed by US hegemonic power, had the intended effect of constraining combatant behavior on all sides—both rebel and state—to include Western militaries that abided by and worked to uphold these international laws.[1] These features were defining until China, Russia, and Iran, with their proxy forces, emerged to directly and indirectly challenge myriad Western-inspired juridical arrangements. Their power and interests became constraints that combatants had to take seriously. The following details how the unipolar moment and the contingent decisions of states and nonstate actors shaped the subsequent evolution of battlespaces.

The Unipolar Moment and Battlespaces

The West became the geopolitical axis of an immense concentration of ideational, economic, political, and military power. All other states and nonstate actors had to adjust. Geopolitical changes induced a cascade of domestic changes in states that relied on Cold War–era resources and political support from superpowers to maintain their regimes. Suddenly they were forced to implement domestic political and economic reforms that included at a minimum a veneer of multiparty democracy and some semblance of neoliberal economic policies. As the former Soviet Union and its satellite states adopted multiparty systems, authoritarian states in other parts of the world had to change too. China, after suppressing the 1989 Tiananmen Square pro-democracy movement, consolidated domestic state control while managing a radical economic opening to the Western-oriented global economy. Domestic economic reforms in many countries mirrored (and often copied verbatim) World Bank and International Monetary Fund policy recommendations. Multilateral institutions focused on expanding and deepening economic growth that facilitated the rapid integration of markets and movements of people across borders. These sweeping economic changes came to be commonly referred to as globalization. Globalization altered economic incentives and stepped up the pace of technological innovations, turning advanced levels of basic and applied scientific knowledge into day-to-day, taken-for-granted realities.

There was also a fundamental shift in ideational foundations. "Liberalism" and "liberal internationalism" were possible as hegemonic ideational concepts because of the concentration of political and economic power in the West, steady pace of globalization, waves of democratization, and end of communism and socialism as viable ideational alternatives to capitalism and democracy. This liberal ideational hegemony was manifest in terms of geopolitical and political realities: democratizations, liberal economic reforms, and multilateral institutions like the World Trade Organization, juridical bodies like the International

Criminal Court (ICC), and multitudes of mostly Western-funded nongovernmental organizations (NGOs) and intergovernmental organizations (IGOs). Some of these organizations and institutions existed before the end of the Cold War, but they had nothing like the global reach and capacity to intrude in the domestic affairs of sovereign states that they acquired after the Soviet collapse in 1991.

Put simply, it was irrelevant whether countries, leaders, and people agreed or not with the Western creation of transnational judicial, political, and economic realities. Even though these realities by design restricted choices available to states, they were unavoidable. States simply had to rationalize and contend with them. Ideas turned into institutional realities also shape behaviors. This effect is seen in a novel discourse that years later one hears not only in Western capitals but also in unexpected corners when nonstate armed actors and even tin-pot dictators insist that they are defenders of human rights, women's rights, minority rights, and, under more intense pressure, LGBTQ+ rights, and the like.

This ideational intrusion into domestic affairs of otherwise sovereign states had been seen before, such as in the nineteenth-century practice of European great powers forcing upon the Ottoman Empire formal rights for ethnic and religious minorities. But the great powers' objective was strategic. It was to promote political stability in volatile regions. Ideational considerations were present, and indeed demands, such as for economic and political equality for Jews, went beyond what their own laws provided.[2] Those measures resembled post–Cold War practices in that extreme power asymmetries allowed the strong to intrude into the affairs of the weak. But nineteenth-century practice was tied to definite strategic ends and lacked the intensity of conviction held by the post–Cold War political establishments in the United States and many European countries that their ideational framework was truly universal, self-evident, and, most importantly, welcome in the societies that they proposed to restructure.

Ideas become manifest realities in two ways: genuine conviction or imposition. Individuals, communities, or states identify an intrinsic value inherent in the idea. The Helsinki Accords, signed by thirty-five states in 1975—essentially the entire Western and Soviet zones of vital interests—illustrate the power of conviction to change the ideational element of the GRINS framework. Soviet and Eastern Bloc governments saw the accords as codifying a status quo with Western acknowledgment of post-WWII borders—a final and irreversible division of Germany into separate eastern and western states—and mutual promises of nonintervention into each other's domestic affairs. Eastern Bloc officials dismissed the accords' statements about human rights and freedom of information as unenforceable concessions for political convenience. But dissident groups soon formed in Eastern Bloc states, including in the Soviet Union itself, to monitor and report violations. Now they could claim that their own governments had made commitments to these values, and this gave these groups the right to challenge government behavior. They became part of a transnational network as Helsinki Watch (later Human Rights Watch), formed in 1978 to monitor Soviet compliance with the accords.[3]

This "Helsinki effect" demonstrates the impact of a change in conviction about ideas on the GRINS geostrategic element. The ideas of prominent dissidents like Czechoslovakia's Václav Havel (a founder of the Charter 77 monitoring group and future president of the Czech Republic) were seen in a new light. Havel began his 1978 essay *The Power of the Powerless* with the statement "A specter is haunting Eastern Europe: the specter of what in the West is called 'dissent.'"[4] He advised that one "need not accept the lie" and that it was possible to break the rules of the game to "live with the truth." Soviet dissident Nathan Sharansky (future Israeli politician), imprisoned for his role in setting up the Moscow Helsinki Group in 1976, wrote in his 1988 memoir *Fear No Evil* that he relied on the Psalms of David to guide his inner conviction to separate the truth of freedom of conscience from the lies of Soviet ideology.[5]

Global institutions and networks like Human Rights Watch shape the behaviors of states and people. Their capacities to do so also reflect power relations at the time of their creation. Havel and others relied on the force of their ideas, but undeniably this process benefited from the ideational, economic, and political power concentrated in the West. The West defined and devised the many tiers of multilateral institutions that became genuine geopolitical realities that states had to contend with. Once the Soviet Union was no longer a military threat, those that challenged these new realities faced the wrath of Western, specifically American, military might. Novel battlespaces emerged at this confluence, at the nexus of Western decisions and adversaries' choices.

Part of the ideational movement that created transnational institutional and juridical realties did not envision force as an acceptable means of creating political effects. But those who defied changes saw the advance of some ideas in terms of the West imposing a novel geopolitical reality. Russia's future president Vladimir Putin wrote of his recognition of this geopolitical reality behind the force of ideas. While facing demonstrators outside the Dresden headquarters of the East German Stasi (secret police), Putin described how he called the headquarters of a Soviet Red Army tank unit to ask for protection. The response was "We cannot do anything without orders from Moscow. And Moscow is silent."[6] Unlike during the Cold War, now any state or nonstate protagonist using violence to create effects had to navigate not two superpowers but a complex, layered geopolitical reality with a single military power—the United States—and economic power in Japan and the European Union. Yet the United States as the sole superpower was unencumbered by this novel geopolitical reality.

For a moment, it was as if the end of the Cold War had robbed the United States and its allies of their animating military raison d'être as the enemy had suddenly ceased to exist. The US armed forces emerged with unrivaled strategic primacy and conventional combat primacy across the land, sea, and air domains, with incipient cyber and space capabilities. The United States also retained a qualitative and quantitative edge in its strategic nuclear arsenal with its state-of-the-art submarine fleet and advanced supersonic and subsonic stealth delivery capabilities.[7] American strategic primacy was made possible at the confluence of military organizational reforms, steady incorporation of new technology, an alliance system that remained intact, and a robust military architecture that spanned the globe with over 800 military bases (to include airbases, ports, prepositioned logistical nodes, etc.), and a real estate portfolio valued at over $1 trillion.[8] This global security architecture, coupled with the combatant command structure of its armed forces, provided the United States with unrivaled combat primacy. Militarily, the world had reached a unipolar moment. That meant that the United States and whatever coalition of countries it could assemble could agree on a desired political end state and rely on its military to create desired effects with force, exempt from institutional realities that the United States and its allies were devising. The unipolar moment was complete.

Battlespaces are also about agency. US and various coalition uses of force within the altered global reality, such as in Somalia, the Balkans, Iraq, Afghanistan, and Libya, help to illustrate the trend lines that eventually formed the contours and character of emerging battlespaces.

Strategic Narcissism and Warfare in the Unipolar Era

The end of the Cold War was transformative, but two geopolitical conditions remained constant. The logic of nuclear deterrence held, and it made a conventional war between major powers unlikely. The post-WWII global consensus remained that existing states would retain their de jure sovereignty no matter how weak or fictitious the capacities of their governments to maintain domestic order. That consensus was reflected in the rarity of interstate war more generally. Everything else was in flux.

The Cold War, with its bipolarity, sustained a structural rigidity. That rigidity shaped the choices protagonists faced in their use of violence to create political effects. In the unipolar moment, the United States emerged with combat primacy; for the most part it had the rare luxury of being able to do what it wanted without regard to the sorts of strategic-balance calculations that were integral to any operation during the Cold War. This overwhelming US primacy created the conditions that enabled global institutions, courts, norms, and conventions, including ones that were critical of US policies and US power itself, to shape the behavior of international actors.

The GRINS realities applied in this unipolar context, as they had before that time and as they do now. But the unipolar geopolitical element gave free reign to a strain of strategic narcissism in the pursuit of a moral vision in international politics. Coined in the midst of the Cold War by international relations theorist Hans Morgenthau, the term "strategic narcissism" refers to an American tendency to view the world only in relation to the United States and a desire to impose on other countries a political system and moral code created in an American image. Morgenthau thought that this was dangerous because, like an individual's narcissism, it perilously ignored the objective conditions that affected the individual's fate.[9] This is a term that H. R. McMaster, President Donald Trump's national security advisor from 2017 to 2018, used in his critique of more recent policies.[10] As it turned out, the Cold War contest between the United States and the Soviet Union forced a mutual understanding of the limits of force. But as the Soviet Union collapsed, considerations about the utility and limits of force underwent radical revision.

There was a near consensus in the American foreign policy establishment—a group that includes government officials but also prominent academics, journalists, and commentators—that American values of democratic governance and openness to global markets really were universal. The mass protests that played pivotal roles in the collapse of Soviet power appeared to confirm that conclusion, as did images such as student protestors carrying a model of the Statue of Liberty in Beijing's Tiananmen Square in 1989. The American political scientist Francis Fukuyama, a deputy director of policy planning at the Department of State and a RAND Corporation analyst, wrote in that year, "The triumph of the West, of the Western *idea*, is evident first of all in the total exhaustion of viable systemic alternatives to Western liberalism . . . the end-point of mankind's ideological evolution and the universalization of Western democracy." Though recognizing that religious extremism and ethnic conflict remained real possibilities, Fukuyama argued, "It matters very little what strange thoughts occur to people in Albania or Burkina Faso, for we are interested in what one could in some sense call the common ideological heritage of mankind."[11]

In hindsight, mass protests against Soviet interference in the domestic affairs of countries in Eastern Europe and dissent within the constituent republics of the Soviet Union from the late 1980s look more like assertions of nationalism and self-determination than yearnings to trade one superpower's ideas for another's about the best way to run one's own country. The events of 1989 are better seen as part of the broader historical process of self-determination than as emulation of a victorious hegemon. Those events are easier to understand in terms of the breakup of multiethnic empires and the rise of national states that gained traction in Europe from the French Revolution. Liberal ideas had genuine appeal to many protestors, not least because initially the US and Western European governments were in no position to insist on imposing these ideas. This absence of domination made it easier for protestors who challenged their own governments to side with the US government and foreign policy establishment.

This brief honeymoon ended as the reality of unipolar hegemony came into conflict with this element of self-determination in the relationship. Politics returned soon enough and in fact had never gone away. But the immense military and economic power of the United States that made the moral activism of the US and broader Western foreign policy establishment possible also buffered it from the consequences of its choices—thus enabling a more serious case of strategic narcissism than Morgenthau imagined. The dilemma was the reverse of the Cold War strategic paradox: the United States now was much less bound in how it could use force. But the consistent overestimation of its moral prestige and consistent underestimation of the resolve of those bent on pursuing their own aims required open-ended commitments of force to save them from themselves. There was, as it quicky transpired, a need for constant revision of the ends to which force was used, which aroused skepticism among the domestic publics of the countries that fought these wars.

Wars of State Failure

Civil wars within the very weak states of the global periphery (such as in Somalia, Liberia, Sierra Leone, and the Democratic Republic of the Congo) and the collapse of multinational federations in the Soviet Bloc (the former Yugoslavia and the Soviet Union itself) arose in the early 1990s as the main arenas for warfare. These new wars appeared with the end of the civil wars that were the explicit extension of the interstitial attrition strategy of global hegemons during the Cold War, such as in Angola, Mozambique, and Nicaragua.[12] The patronage networks that held these places together in lieu of formal institutions found themselves unable to adjust their domestic regimes to the changed global political and economic realities.[13]

The collapse of the Somali state was complete by late January 1991 when President Mohamed Siad Barre and a few core political supporters fled the capital in a tank after a stop at the central bank to pick up what remained of the country's reserves.[14] Four separate rebel groups converged on the city as mass killings ensued. While these massive abuses of human rights did not immediately trigger Western intervention, the onset of famine prompted the establishment of the United Nations Operation in Somalia (UNOSOM) in April 1992. The peacekeeping operation's mandate was to monitor a shaky cease-fire and protect convoys to deliver famine-relief supplies. As the security situation deteriorated, the UN Security Council expanded UNOSOM's mandate to include the Unified Task Force to secure populated areas and organize the distribution of relief supplies.

The multinational armed intervention in Somalia reflected a pattern in the 1990s of UN-sponsored multinational armed interventions in multisided civil wars amid state collapse in sub-Saharan Africa (Liberia, Sierra Leone, the Central African Republic, and Congo), and, as examined at greater length later in this chapter, the former Yugoslavia. Since these UN interventions required the consent of all five permanent members of the UN Security Council (the United States, United Kingdom, France, China, and Russia), these armed "peacekeeping" missions would have been highly unlikely during the Cold War geopolitical stalemates between the United States and the Soviet Union. China's consent reflected its inward focus and still peripheral position in global geopolitics. The United States' overwhelming political and military power relative to the rest gave the American foreign policy establishment free reign to incorporate the legitimacy of UN mandates and the coercive force of the foreign military contingents that made up these operations to support its vision of what a US-dominated global order should look like.

Even though the US intervention in Somalia in 1993 claimed the legitimacy of a UN Security Council mandate, from the perspectives of many Somalis with Fukuyama's "strange thoughts" in their heads, the uninvited appearance of foreign soldiers in the middle of their war made the United States their enemy. The passions of nationalism and self-determination that worked in favor of American power in Eastern Europe now turned against it. The *Black Hawk Down* battle in Mogadishu on October 3–4, 1993, that left twenty-one American soldiers (and one Malaysian and one Pakistani soldier) dead forced hard questions on Americans: What were the objectives of the intervention? What was the desired end state in terms of defined US interests? Like the Cold War–era direct engagements of US forces, the intervention in Somalia did not seem to generate strategic returns commensurate with the investment of American resources and lives.[15] Almost two generations later, a Western-backed African Union force remains in the country as US Special Operations Forces train Somali fighters and chase Islamist rebels.[16]

The intervention in Somalia reflected the changed ideational context in which the United States and Europe claimed the right of the "international community" to intervene in the domestic affairs of states in which governments (or in Somalia, local warlords) oppressed civilians. UN Security Council Resolution 688, adopted on April 5, 1991, placed Iraq's Kurdish population under international humanitarian protection and emerged as a key precedent that established the principle of intervention to save civilians from their own governments. The United States, as the third-party intervener, used its overwhelming power to impose order, but doing so inevitably generated casualties and collateral damage. The United States was an occupier, no matter how well intentioned.

If the United States picks a domestic partner and turns it into the winner, victory will come, as in all wars, at the cost of bloodshed, which will bring international public opprobrium. The political scientist Roy Licklider found that about three-quarters of all civil wars end when one side wins a decisive victory over the others. Of the quarter that conclude with a negotiated settlement, half of those return to wars that usually end when one side wins a decisive victory.[17] But in an idealized—a narcissistic—conception of war, no one is willing to assume the risk of making this utilitarian decision to support a likely winner, however unpleasant in the short-term, even though it may generate the greater good in the long term. The prevailing peacebuilder intends to intervene to force all armed groups to take part in a negotiated settlement to share power. Politics is discussed, but all too often, political realities are ignored.[18]

When political realities cannot be ignored, tactical retreat and handoff to local proxies, humanitarian NGOs, and a complex array of international institutions ensues. They are left to take over the "state-building" task after this pragmatic exit. Facing the intractability and near-zero strategic return on its military investment, the United States beat a hasty retreat and turned Somalia over to international organizations and contractors, making it a ward of the international community. The United Nations and international organizations have engaged in a fantasy state-building enterprise from Kenya and Djibouti since 1992 at a cost of well over $55 billion.[19] This fake Somali state-building enterprise has been turned into a very lucrative external patronage network that directly funds local armed actors who walk in the guise of politicians, IGO advocates, NGO leaders, warlords become "peacelords," and transnational businessmen in collaboration with donors.[20]

A defining lesson of Somalia was that while state boundaries remain inviolable, at least in juridical terms, the nature of the domestic regime plays a defining role in shaping the character of civil wars.[21] And the driving force, the ideational justification for the war, can take myriad context-specific, historically rooted forms (e.g., grievance, greed, ethnic, religious, urban, rural, etc.).[22] The nature of military organizations and scientific knowledge was directly shaped by the nature of the regime and the socioeconomic context. Rudimentary technological innovations and the use of mass casualty produced weapons with no regard for collateral damage. This combination of factors generated forms of warfare on the ground that bedeviled Western interveners and created effects that continue to generate calls for international actions to rescue civilian victims.

State failure in West Africa, specifically Liberia and Sierra Leone, had common themes. They were "state-nations" that had become personalistic regimes with barely a veneer of formal institutions and rulers who leveraged Cold War political alignments for resources. In lieu of domestic institutions, elite patronage networks, defined along ethnic, clan, religious, economic, and political lines, constituted governance networks with real power. In this light, it was not a surprise that two of sub-Saharan Africa's three largest recipients of US assistance—Liberia and Somalia—promptly collapsed when that assistance was cut off once their human rights abuses and corruption dominated American considerations. The collapse of the third country—Congo—took a few years, but it hit hard, generating a conflict that destabilized the midsection of the entire continent and has hosted multiple UN peacekeeping operations since 1999.

In an altered geopolitical context, when internal dissention turned into civil war, without proper domestic security institutions, insurgencies (intrastate war) became a violent reflection of the preceding regimes. That is, civil wars became contests between former patronage networks, although this time the elite contestation had become militarized.[23] The logic of violence when it is a militarized contest between patronage networks is difficult to interpret from afar, when even people in the local neighborhood find the situation too fluid. Seen from the West, all the protagonists seemed the same; everyone appeared to be killing each other for no rhyme or reason. Inability to contextualize and accurately interpret micro-level dynamics led to violence often being interpreted as senseless. Who should be shooting whom was shaped by context- and domain-specific knowledge, and whereas local combatants were mostly aware of this nuance, many external observers saw no logic.[24] International observers, unable to understand the localized logic, interpreted it in terms of insanity, though on the streets, the supposedly insane reality was generated by local big men who were exceptionally sane in pursuing a logic of violence based on appropriateness, rooted in their home turf, that they knew all too well.[25]

Local armed actors relied on rudimentary technology that reflected acute domain-specific invention in its technology-human interface. Combatants could rely on versatile, dependable, and cheap weapons systems: light machine guns that they could shoot from the hip. To many Western observers, all of these combatants look the same, such as when one is concerned foremost about the rights of the child or the gender balance of intervening forces. The narcissistic rejection (and ignorance) of political realities spawned critiques by Western publics that paid attention, such as British actor and television host Russell Brand's 2010 song *African Child (Trapped in Me)*.[26] It verges on understatement to note these conflicts lacked hygienic clarity sought by lawyers, human rights activists, and Western military officials increasingly taking a legalistic approach to target selection and apprehension for international tribunals.

In 1994, another civil war quickly descended into the Rwandan genocide as that country's rulers used the conflict as a pretext to put their genocidal fantasies into operation.[27] Images of genocide—officially deemed possible acts of genocide but never genocide—streamed into Western households by way of satellite TV. The United States and its allies faced a dilemma. Ideationally it would be morally inconsistent to abrogate the responsibility to act, especially considering such evident suffering of the innocent. Only a year prior, well-intentioned Americans entered someone else's war in Somalia to alleviate suffering only to be killed by the locals they were there to protect. When the right and the smart do not align, politics is about a trade-off between the right use and the smart application of power. Why should US service members die in some distant land that the average American saw as having no direct relevance or posing no direct threat to national interests? American leadership decided against intervening.

The United States and likeminded officials in other countries, primarily in Western Europe, instead empowered international organizations to deal with the humanitarian suffering, effectively outsourcing the effort. This decision, buttressed by billions of dollars, created its own industry. Centered on new concepts of peacemaking, peacebuilding, and peacekeeping, the humanitarian peace builder industry developed their own logic of politics, despite little evidence to suggest it was effective in the nation- and state-building processes. The decision also turned the US Agency for International Development into a contracting agency, where, unbeknownst to taxpayers and with little to no accountability, tens of billions of dollars were dispersed to contracting agencies with very little to show for the outlays. Contracting out services has perverse dynamics in civil wars: astute combatants use violence as a part of the negotiations process with NGOs and IGOs that bring the peacebuilding gravy train of resources needed for sustaining patronage networks.[28] Simply put, for some combatants, war became less about regime change or political grievance, and the battlespace transformed into competition for international aid.[29]

Not using the US military to stop these catastrophes generated moral outrage among some intellectuals. This righteous rage spawned a new ideational movement that crept onto college campuses and into think tanks and halls of power, eventually becoming its own industry. The enterprise of saving people from themselves took root in the late 1990s and matured as a twenty-first-century civilizing mission. Its advocates tried to pair humanitarian causes with the need for military interventions, turning a humanistic impulse into a geopolitical principle. In 2001, the UN-appointed International Commission on Intervention and Sovereignty codified a shift in focus from the right to intervene to the responsibility to protect (R2P).[30]

This interventionist notion that people must be emancipated from the violence of their own governments found support in the US government. Anne-Marie Slaughter, the former dean of Princeton's Woodrow Wilson

School of Public and International Affairs, urged US intervention in Libya while she was director of policy planning at the State Department. She later wrote that "it clearly can be in the US and the West's strategic interest to help social revolutions fight for the values we espouse and proclaim" and that Libya was on the right path with a "constitutional charter that is impressive in scope, aspirations and detail—including 37 articles on rights, freedoms and governance arrangements."[31] She went on to argue for US intervention in Syria as public order in Libya collapsed amid factional violence, presenting proposals with no clear end state.[32] Samantha Power, a Barack Obama administration ambassador to the United Nations, also advocated for armed intervention in Libya.[33] At a congressional hearing in 2021 for her nomination as head of the US Agency for International Development, she defended her support for what Obama called "the worst mistake" of his presidency on the grounds that it protected civilians from harm, though "the fallout after the intervention—the centrifugal forces—have been incredibly difficult to manage and above all, hard on the Libyan people," she added.

The critical element of these foreign policy failures was that US military supremacy was a necessary condition for their commission; yet each failure undermined the US strategic position in the world.[34] This open-ended pursuit of supposedly universal principles of governance models predicated on American ideals eschewed national interests or balance-of-power concerns and approached ways and means with a disregard for political contexts on the ground. This strategic narcissism, manifest in regime change in targeted countries, put other regimes on notice that they needed to act (as in the case of Iran's and North Korea's nuclear weapons programs) to prevent this doctrine's application to their regimes.

The Yugoslav Wars in the Balkans: Strategic Focus

Wars of state failure and state transformation also appeared at the edge of Europe. The conflict processes on the ground in the Balkans and the gratuitous nature of violence were similar to what was found in the wars in West Africa. But the United States and Europe responded differently to wars breaking out in a disintegrating Yugoslavia (1991–2001). The US and North Atlantic Treaty Organization (NATO) intervention in the Yugoslav wars showed that military intervention for defined end states could manage limited success but at the cost of long-term commitments to police the aftermath.

Collapsing empires and superstates create opportunities for oppressed communities to fight for national self-determination. The Cold War froze in place power relationships with political structures that were often incongruent with social realities on the ground. This was a pronounced reality in states behind the Iron Curtain. Like the Soviet Union, Yugoslavia was organized along national lines, with federal borders mostly congruent with the ethnic dispensation on the ground.

Slovenia and Croatia broke away first and were recognized by European states, ending the Yugoslavian Federation. With the federation now doomed as European governments recognized the breakaway states, the political realities on the ground played a much greater role, compared to in conflicts in Africa, in shaping how the US and NATO member governments applied force. The breakup of Yugoslavia was characterized by the gratuitous violence seen in the African civil wars but also presented interveners with armed groups that differed in their organization—which played a defining role in shaping the subsequent peace. The autonomous nature of federal entities meant the war in Yugoslavia had a conventional character with formal armed forces fighting each other. Each side relied on Soviet-era armaments, with assorted proxies doing the dirty work of removing undesirable ethnicities. Like Africa's wars, it was also a war where the effects of globalization, local-global networks, and commercial networks suddenly played a more enhanced role in shaping the protracted character of war, which led some scholars to suggest that this is the beginning of a new kind of war.[35] But the fact that this was a war in Europe forced the United States and its NATO allies to privilege political ends over (very real) moral concerns.

The United States chose to intervene decisively after European peacemaking initiatives failed. It first engaged covertly and created realities on the ground, primarily through supporting Croatian and Bosnian forces to drive Serbian government and ethnic Serb militias out of the areas they occupied, followed by an overt air campaign that forced a peace settlement. Then the United States brought in the United Nations and European Union to formalize and legitimize the peace through NATO soldiers working to keep peace between all parties.

The American approach reflected strategic aims, such as maintaining US dominance in the NATO alliance after the Cold War. The failure of European efforts to resolve the conflict prior to large-scale US engagement from 1993 showed the United States' European allies that the US presence was still necessary to provide security in Europe as the Europeans were constructing the European Union. Though American officials cited human rights concerns, the engagement defined strategic ends and acknowledged the utility that partnerships with even very nasty actors could have in achieving these ends.

In any war, adversaries have a vote. The intervention succeeded, but only because the war was waged by formal actors with formal organizational structures—and in many cases, armed actors opposing the West did so for reasons that seemed irrational to outsiders.[36] Therefore, when the leaders signed the peace deal, there was a peace to keep, as the rank and file acquiesced, as did the people, though the civil war was mostly an elite affair rather than a grassroots one.[37]

The rest of the world, especially Russia, perceived the NATO intervention—which lacked UN authorization—as an abuse of power, even though NATO leaders thought they were doing the right thing. It was perceived by others—Russia and China especially—as American combat primacy translating into a means to do what the United States wanted for the right reasons, defined by them—conventions and norms that the United States in particular was erecting that it expected others to adhere to.[38] The involvement of NATO took place just as it was expanding into former Soviet satellite states, abnegating a verbal guarantee to not do so supposedly given to Soviet leaders by the George H. W. Bush administration at the end of the Cold War.[39]

Finally, the NATO intervention demonstrated superior firepower. It showed how the Joint Force concept could be operationalized in an optimal fashion. Such demonstration of effective power projection also generated concern. The Chinese, Russians, and myriad other regional adversaries in the Middle East took note of the United States' capabilities and low-level flexing of its advanced military might. No power could counter the US and NATO position because these actions in the Balkans generated peace, avoiding a regional conflict spillover into Europe, a sphere of interest that the United States identified as its own.[40] The cessation of West African and Balkan wars gave rise to geopolitical realities that influenced and shaped subsequent battlespaces.

Constructing Geopolitical Realities: Transnational Justice Movements

The years following the end of the Cold War witnessed an ideational phenomenon in the form of proliferating international criminal tribunals and proceedings. They were tailored to the atrocities in Yugoslavia, Rwanda, and Sierra Leone and the protracted proceedings in Cambodia.

The cascade of tribunals and special courts culminated with the signing of the Rome Statute in 1998, creating the International Criminal Court.[41] It was an extension of the broad liberal—in the Kantian sense—ideational mood of the time.[42] This ideational social movement was as much about justice and retribution as about manifesting an ideational outcome and altering power relations by constructing novel geopolitical realities.

Imposing formalized moral logic with juridical tools inevitably generates contradictions when faced with power realties. At the turn of the twenty-first century, the United States was not subjected to the same scales of justice as the rest. Arguably because it and its assorted coalitions were playing the role of arbiter and policeman, the situation was viewed with approval in many US quarters, and its rejection by those to whom it was applied was taken as evidence of their failure to improve themselves. Further, international justice begins at the moment of apprehension; one could cajole and bribe some states to hand over perpetrators, but the most hardened of them all would require special attention. The US government utilized Tier 1 Special Operations Forces to conduct covert operations to bring all those responsible to justice, especially in the Balkans. In other words, turning an ideational enterprise into a juridical reality by way of politics inevitably requires enforcement, and internal contradictions occur in enforcement. This was not lost on anti-Western foes.

Among the many shelves of books and reports dedicated to justice cascades on academics' and activists' bookshelves, it is impossible to find the role of covert operations, a mechanism that made the enterprise possible. Either it is a genuine omission because the scholars were truly unaware, or it is a deliberate omission to maintain ideational and juridical purity or because subjecting the same covert operators that made the initial reality possible to justice later would require a studied ignorance. In either case, this is an imposition of a newly imagined sovereignty. Eventually, these actors would weaponize those moral, ethical, and legal contradictions to generate tactical- and operational-level advantages in their respective battlespaces the moment the West walked into the asymmetric trinity: Afghanistan (2001), Iraq (2003), and Libya (2011).

Asymmetric Conflicts

The United States faced an act of terrorism in the suicide attacks on New York and Washington, DC, on September 11, 2001, conducted by numerous men with conviction consumed by a totalizing ideology. Immediately, the United States and its allies once again confronted the puzzle that Niccolò Machiavelli tied into a Gordian knot. If one is faced with two ends, equally ultimate, equally sacred, and they contradict each other without the possibility of rational arbitration, to untie the knot is to live and let live. And that is the ideational basis of pluralism in liberal democracies. The ideas are in constant tension; people agree to disagree but do not disagree on the disagreements since the political institutions create processes of arbitration in the form of elections. In the face of a violent ideological movement with a millenarian vision, with no room for compromise, untying the Gordian knot is not an option. It must be cut. In that moment, violence once more becomes the final arbiter of what is sacred and what is ultimate.

The NATO alliance invoked Article 5 and thus began a protracted conflict in Afghanistan that grew indirectly to encompass Iraq, as well as myriad other nonwar wars in the Pacific, the Sahel, and the Horn of Africa. Though NATO intervened in Libya on very different premises, these conflicts were broadly characterized as variations of asymmetric irregular conflicts.

The doctrinal definition of *irregular warfare* is like recent definitions of *insurgency*. It refers to a conflict where nonstate armed actors fight a legally constituted state for control and support of a population. *Asymmetric conflicts* refer to instances where the conflict is characterized by warring parties with vast asymmetries of power (resources), and the concept is also stretched to include asymmetries in strategies, operational approaches, and tactics. Western interventions were asymmetric in every dimension. The West attempted to impose a political structure that mirrored its own institutions. In pursuit of that end state, the West implemented a strategy that partners found to be incongruous with their own interests, yet grudgingly accepted. The Global War on Terror era created a globalist counterinsurgency and counterterrorism campaign in pursuit of destroying various insurgencies and terrorist groupings that opposed US objectives.[43]

Operationally and tactically, the United States and its partners approached the conflict on its own terms, irrespective of their utility in creating effects on the ground. Ideally, to be able to attack an adversary's strategic centers of gravity is to gain the initiative in war. Facing an adversary whose policies, strategies, operational art, and tactics remain incongruous both with each other and with the realities on the ground, weaker adversaries find maneuver space in those incongruities at all levels of war. An improvised explosive device here, an assassination of a political leader there, intimidation of schoolteachers somewhere else—though tactical in their manifestation, these seemingly discrete acts by weaker adversaries amount to an attack on the broader strategy of the stronger adversary, when the stronger adversary owns the battlespace, fighting someone else's fight.

The Backyard of Empires: Afghanistan

Throughout history, when a great power is subject to perfidy, dishonored, challenged, or attacked without provocation, swift retribution inevitably follows. The nature of great power politics, bound up with security of the realm, sovereignty, territorial integrity, and intangible realities of honor, credibility, and influence, necessitates retribution. A classic example is British general Robert Napier's 1868 punitive expedition in Abyssinia (modern-day Ethiopia) to rescue British missionaries and officials imprisoned by Emperor Tewodros II. In the most expensive military expedition ever undertaken for honor, the British sent a force of 13,000 troops (mainly Indian sepoys) that included over 20,000 attending camp supporters.

The British had to make an example of the Abyssinian emperor to demonstrate that unmitigated insolence could not stand. Described as a "punitive expedition" by British politicians and military leaders, they saved the British hostages and ousted Emperor Tewodros. They destroyed and looted his palace in Magdala, then promptly returned home as Abyssinia devolved into internecine fighting.[44] What form and objective would the US and NATO intervention take in Afghanistan?

Afghanistan, like Switzerland, is a backyard, not a graveyard, of empires. Some countries came into being because their powerful neighbors wanted them to exist, such as Belgium, which became a buffer state. Other countries came into existence as a by-product of the complicated processes of state formation, war making, and imperial enterprises. Some geographic spaces and people, by virtue of a complicated human geography, were sometimes not worth the cost of conquering or annexing.

In an era of nation-states, Switzerland consolidated its many cantons, which were the appendages of empires, into a multilingual, multiethnic, multisectarian state-nation. Swiss mercenary warlords of the day eventually came to a working arrangement to fight others but not each other. Switzerland's mercenary heritage is still noticeable with every male required to serve in the military and issued a firearm by the government. However, Switzerland as a nation ceases to exist the moment

Germans, Italians, and the French meet on a football pitch. Afghanistan is the Switzerland of central Asia, a multiethnic, multilingual, multisectarian state-nation with challenging terrain far more diverse and breathtaking than Switzerland's. In a near-identical process, Afghanistan emerged as the backyard of multiple empires when the empires made a virtue out of necessity by coming to a consensus on the current boundaries of Afghanistan.

Neighborhoods matter in geopolitics as in life. Switzerland, by virtue of being in Europe, experienced modernity and benefited from it, while Afghanistan, given its neighborhood, got the short end of the stick in terms of the ideational, economic, military, and political benefits of modernity. Neighborhood, sociocultural endowments, and the heavy burden of ideas in terms of religion, coupled with the cruelty of history, made it a harsh country with normalized ultraconservatism. Numerous attempts, beginning in the late 1920s, to bring some modicum of modernity, as Afghan leaders wanted to emulate Atatürk's Turkey, consistently failed due to the fractured power of the state and the strength of tribal conservativism.[45]

Being the backyard of empires, Afghanistan became one of the bloodiest battlegrounds of interstitial warfare between the United States and the Soviet Union during the Cold War. This led to a form of arrested development in Afghanistan, where the modern exists alongside the medieval. Consequently, American and NATO soldiers still marvel at the sight of Afghans living in mud huts and caves while using cellphones. This is the sort of observation made by British antihero Harry Flashman while serving in Afghanistan 1842.[46] It is not that Afghanistan is the same as in Flashman's day, unreached by modernity—hardly. But modernity entails an inevitable altering of existing power relations, and in Afghanistan externally induced incentives have made countermodernist currents most prominent. The anachronistic emirate of the Taliban was animated by a purist ideational vision that was, on the one hand, as organically Afghan as the sublime Afghan hash. On the other it was as foreign, alien, and incongruous to Afghanistan as Taliban fighters riding around in a Toyota Hilux "technical" (typically a truck with a heavy machine gun bolted down in the bed), enforcing their ban on music and women wearing makeup. American and allied troops stumbled into this context.

The US-led intervention into Afghanistan would be about punitive action, but the ideational context of the world had changed. Unlike the British expedition to Abyssinia, it could not just be about killing some people to make an indelible point and return home. The puzzle was that the West had won the Cold War and had to maintain the veneer of a proper world order, especially with the use of violence for a reasonably justified purpose besides justice and retribution. Many public intellectuals in the United States felt their country had reached the pinnacle of human existence; yet the Taliban were the complete ideational opposite of the "Davos Man." What could be done to a regime that had no interest in joining the international community? Could this intervention finally modernize Afghanistan?

Humanitarian interventions, even if done unilaterally, worked in the Balkans in the 1990s (e.g., Bosnia, Kosovo), with winners and losers. The presence of Western peacekeepers in the Balkans—the Stabilization Force in Bosnia and Herzegovina (1997–2004), which later became the European Union Force Bosnia and Herzegovina (2004–present), and also the Kosovo Force (1999–present)—could (more or less) put fractured countries back together. Not everyone wanted these peacekeepers, but they were enough to temper hostilities by many of the aggrieved, and their continued presence and the pressure valve of migration and remittances fostered stability and a form of democracy. As noted above, violence, properly applied with a clearly defined political end and domestic buy-in arrived at with a mix of coercion, collusion, co-optation, and genuine cooperation, could generate positive outcomes as seen in the Balkans. Could similar mediating actions work in Afghanistan?

The United States intervened in punitive fashion while allowing NATO to guide Afghanistan's political administration. The military intervention was supposed to support the political objective of an Afghan state in the image of the West. This translated into a civilian army of people landing in Afghanistan from international organizations, multilateral agencies, NGOs, and research institutes, each with the best of intentions and its own vision of modernizing the country.[47]

Different visions meant different things to different people. Americans would teach Afghans how to be militarily effective by abiding by Western humanitarian norms. Italians would train the Afghans to build effective police forces. The Dutch would teach the Afghan security institutions how to unionize and engage in collective bargaining. Meanwhile the erstwhile US ally Pakistan would choose to harbor Usama Bin Laden (UBL) while facilitating the rearming and reorganization of the Taliban as its proxy.[48] The Afghan government in Kabul, animated by regime-survival logic, turned the enterprise into a multiethnic patronage arrangement, whereby myriad power brokers (to include their families, cousins, and friends) all got to be a part of the central government.

American and European taxpayers underwrote the Afghan state-building project (2001–present). Those persecuted most (i.e., people from different ethnic, linguistic, and sectarian backgrounds) by the Taliban were thrilled that modernity (or prospects of it) would free them from oppression. Many Afghans living in Herāt and Mazār-i-Sharīf have witnessed the benefits most. Yet, despite the best intentions of Western interveners and the support of a cross section of Afghanistan's population, NATO ended up in the middle of someone else's civil war. Less than two decades later, military parody newspaper articles were identical to the supposed reality that American generals were briefing.[49] Yes, the civil war continues, but parts of Afghanistan remain quantum leaps ahead of where they were in 2001.

When the United States intervened in Afghanistan, the Taliban, primarily a Pashtun phenomenon, with the support of its Arab jihadist allies and Pakistani minders, controlled most of the country, except for the anti-Taliban militias of the Northern Alliance in the Badakhshan province. First formed in 1996, after the Taliban seized control of Kabul after the collapse of the Soviet puppet government, the Northern Alliance consisted of militia leaders representing Tajiks and Uzbeks to include some moderate Pashto leaders who opposed the Taliban. They were already on the defensive, facing possible annihilation, when the United States intervened.[50] With American airpower, the Northern Alliance was a reliable ground army that helped expel the Taliban in two months at the end of the 2001.

As Afghanistan transitioned into 2002, Tajik, Uzbek, and Hazara leaders had the most at stake in terms of what the country would become. After routing the Taliban and after many *loya jirga* (legal assembly) theatrics during the 2001 Bonn Conference, the United States and NATO put in place an impotent Pashto, President Hamid Karzai, with the American military becoming the de facto warlord militia for Karzai. Since then, political leadership and militia commanders from the Tajik, Uzbek, and Hazara have supported US efforts—as most Afghan National Army soldiers come from those ethnic groups. Non-Pashto groups essentially fight to uphold a Pashto warlord (i.e., the Afghan president, sometimes better described as the "mayor of Kabul") who cannot create an effective militia/army or gain control of Pashto regions, all despite Pashto patronage being doled out from Kabul. Although there are disagreements, Tajiks, Uzbeks, and the Hazara do not fight the central state; nor do they fight each other.

The civil war in Afghanistan is simply a civil war between rival tribes within the Pashto. The Taliban—which was hosted, secured, and retrained by Pakistan—fights on so that a proper Pashto warlord can rule Afghanistan. Pakistan played the Afghan situation in its favor. The

84 Old and New Battlespaces

Pakistan government squeezed billions of dollars out of the United States to permit transit and supply lines through Pakistan while passively allowing UBL to live in Abbottābad, in front of the Pakistani version of West Point. Unsurprisingly then, the Pakistani prime minister, Imran Khan, and many others refer to UBL as a "martyr."[51] Pakistan plays the most essential role in the resilience of the Afghan insurgency, getting rich off Americans and NATO and spoiling the prospects of Afghan peace. On the theme of opportunistically bleeding America on the cheap, Iran provides tactical support (e.g., training, weapons, ammunition, etc.), and Russia offers military aid and monetary incentives to Taliban fighters to kill American troops (i.e., bounties). There is no clear end state but rather constantly shifting military end states, and adversaries, friends, and assorted bandits exploit the incongruities. The character of this war and the logic of violence parallel those of the Iraq War (2003–present).

Into a Hollowed State: The Invasion of Iraq

The casus belli for the invasion of Iraq will forever be contested and argued over.[52] While faulty intelligence and real concerns about weapons of mass destruction (WMDs) played a defining role, the ideational shift that took place immediately following the 9/11 terrorist attacks played a decisive role. The ideational shift also explains why most lawmakers, almost all public intellectuals, and over 80 percent of the American population supported the invasion and saw WMDs as a necessary cause of action.

The terrorist experience generated an ideational reality in the United States—and soon elsewhere in Europe—that in an ever-integrating world, terrorism can indeed be an easier means of sowing societal discord. Multiethnic states were especially vulnerable to terrorism since acts of terror pit the innocent against the innocent. When people start viewing each other with suspicion, the sense of civil harmony that characterizes civil societies in liberal cultures of the West weakens from within, and they lose their internal tolerance and cohesiveness. It was also becoming slowly apparent that acts of terrorism, while seemingly isolated, reflected an outcome of a broader, deeply rooted ideational phenomenon built on an extremist version of Islam, with genuine sympathizers who identified with the broad contours of this millenarian vision. The emboldened rhetoric of the terrorist ideologues, their sympathizers, and their surrogates scattered in multiple countries, including in the West, also suggested that this purist, intolerant vision of Islam was a deliberate countercurrent that was aimed at both the liberal pluralist cultures of the West and liberal democratic activists in Muslim countries. These extremists posed a direct threat by way of terrorism and an indirect threat in terms of second- and third-order effects that acts of terrorism create in pluralist democracies. They posed the greatest threat to Muslims in Islamic countries.

If regime types shape the evolution of war in terms of clarification of political end states, then it was inevitable that the objective threat of terrorism, predicated on a millenarian religious vision, would be filtered through political lenses. Consequently, the objective threat of possible WMDs and terrorism was subsumed into politicized rhetoric.[53] Different groups looking at the same objective facts arrived at fundamentally different conclusions based on the political filter through which they viewed the threat. A unique political alliance formed between Cold War–era "war-hawks" and humanitarian "peace-doves," creating "war-doves." These "war-doves" formed the new core of foreign policy elites in the United States and the United Kingdom, yielding a broad enough coalition for an invasion and regime change, all on Western terms: strategic narcissism. The decision aligned clearly with the liberal idealist sentiment at the time and the unbounded sense of Western optimism. And the intervention had domestic procedural legitimacy, with lawmakers authorizing it and military leaders formulating plans to fit the newly established open-ended goals set for them by civilians.

As the Iraq War began in 2003, Iraqis—and most in the Middle East—began observing the dénouement of a war that had been ongoing since 1990.[54] After thirteen years of UN and Western sanctions and a continuous air war, the once highly bureaucratic-centralized Ba'athist state had become a hollow edifice with only a formal veneer. In its place, Iraq constituted a personalized regime, with numerous patronage networks holding it together. The patronage networks were constituted of informal military, ideational (i.e., religious), political (Ba'athist Party membership was used as a social-control mechanism), economic, and localized tribal structures. Unfortunately, the sanctions had shrunk the Iraqi economy such that people had become reliant on the Ba'athist state to a degree that was unthinkable previously, making Iraqis dependent on the one thing they hated the most.

As Americans walked into Iraq with a "coalition of the willing," the edifice unraveled, and the deliberate dismantling of the state by the US administration hollowed out any semblance of formal control instantaneously.[55] Perhaps this was intended, for no one has really explained the de-Ba'athification decision or its logic, but these actions caused the sudden destruction of the Iraqi state.[56] On top of this fractured state, as former military officers point out, a complete breakdown of civilian-military relations in the Pentagon and a lack of military resources in the region created deep and broad ramifications, shaping the unexpected war that followed.[57] In under two years, America and NATO found themselves fighting other people's wars, as they had in the Balkans. But this time would be different.

The US occupation of Iraq fractured the country and turned it into the Catalonia of jihadists. Insurgents were logistically supported at the tactical level by Syria and Iran, while wealthy individuals in mainly Sunni Gulf Arab states supported the jihadists. At the tactical level throughout most of Iraq, it was a *symmetric irregular conflict* in which innumerable Sunni and Shia groups fought each other. It was also an *asymmetric irregular conflict* since everyone wanted to fight the Americans, with the Kurds maintaining a studied distance.

When everyone realized that the Americans were leaving Iraq, Sunni alliances splintered first. And the incentives shifted among the Shia groups to consolidate their gains and retain a guaranteed place at the table by either being political parties with militias or aligning with a larger party with its own militia.

Iran became the greatest beneficiary of the US intervention in Iraq and managed to gain a political, economic, and military stronghold on American coattails. Iran also shifted tactics toward assisting Iraqi Shia clients to consolidate political power, making sure their partners won, and became sectarian kingmakers. The consociational political arrangement that American experts designed—as so many Iraqis point out—in which everyone is expected to win, when faced with political military realities, translated into a constitution where the chosen few will never lose.[58] The same men and a few women who made it, locked into a constitutionally mandated patronage arrangement where they—or their progeny, relatives, or clients—remain power holders, play musical chairs during elections.

Violence, coupled with the vote of the adversaries, meant the United States could turn Iraq into a state with some veneer of stability by the time it left in 2011. This outcome was immensely aided by the sociocultural reality that Iraq, in most Iraqis' lifetimes—before 1990—had been a modern bureaucratic state with among the highest living and health care standards and literacy levels in the Middle East. Despite American missteps that gutted Iraqi bureaucracy, remnants of institutional memory aided in shaping the outcome.

The War of Choice: Libya

The moral logic of violence, as any soldier would point out, is a contradiction in terms. The logical extension of expecting morality and violence to walk in lockstep generates two conclusions that form logical inversions of each other: Damned if you do, damned if you don't. Soldiers, being intimate with the contradiction, do not resolve the irresolvable. At the point of decision, soldiers are expected to own the contradiction and held responsible for subsequent consequences.

Libya forced US and Western policymakers to grapple with strategic-level problems that reflect the same moral conundrums faced by soldiers at the tactical level. The sophisticated articulation of the moral logic of violence, complete with its internal contradictions at the strategic level, is enshrined in the principle of the responsibility to protect, complete with an office and an undersecretary-general in the United Nations. The sociopolitical construction of R2P is a remarkable feat of human ingenuity. This doctrinal assertion claims to resolve the irresolvable by reversing the consequences of the logical inversion. R2P makes it unnecessary to own up to the contradiction and consequent responsibilities at the point of decision if one's intentions are pure. The military intervention in Libya was justified through pure intentions noted earlier. The R2P principle resolved one problem from hell while creating another hell out of the best intentions. Strategic decisionmakers, advocates, and supporters need not be held accountable since their intentions were pure.[59] Unfortunately, R2P adherents and supporters of transnational justice leagues remain blind to the realities of power and hubris that drive their neocivilizing mission. Like all utopian projects, the latest savior impulse, while sincere and pure, is ahistorical and apolitical (in the narcissistic sense of its insensitivity to conditions on the ground). It is predicated on a belief in being able to wield absolute power with self-righteous certitude cloaked in a secular faith. And the secular faith provides certitude and helps one rationalize ironies and contradictions that shape realities on the ground.

Libya had the worst combination of the institutional failures and societal cleavages seen in Afghanistan and Iraq. Leading a coup d'état to take power in 1969, Col. Muammar Qaddafi gutted the incipient institutions that King Idris and the British had put together and relied upon divide-and-conquer strategies between the various tribes. Libya existed as an international pariah until the Western invasions of Afghanistan and Iraq made Qaddafi renounce terrorism and the pursuit of nuclear weapons.

When the 2011 Arab Spring spread to Libya, Qaddafi promised that rebels and protestors would be "hunted down street by street, house by house and wardrobe by wardrobe."[60] British and French leaders were the most concerned in the West, while the US administration was initially cautious about doing anything precisely because the country had many features of Afghanistan and Iraq, and there was little public appetite for a third American ground war. Regardless, humanitarians in the West invoked R2P as a globalized raison d'être. And in that moment, in keeping with the moral logic inherent in R2P, purity of intentions, incontrovertibly argued to the point of clairvoyance without regard to consequences, won the day.

Had the West not led efforts to establish a "no-fly zone," the anti-Qaddafi rebels would have been crushed by Qaddafi's forces, as had happened in previous uprisings—and the West would be damned for inaction.[61] Western air power inevitably led to a "no-drive zone," dictating the survival and success of rebels and the attrition of pro-Qaddafi forces, solving one problem. At the end of the air campaign, airstrikes and intelligence, surveillance, and reconnaissance from Western drones enabled the encirclement of Qaddafi by rebels, leading to his roadside execution. The exuberance of a successful R2P air war to end human rights abuses in Libya did not last long as it became apparent that the West had no post-Qaddafi plan for the country. This created another problem as the victorious rebel leaders did not want any international involvement after their victory. Because of this interest convergence, Libyans could pursue the democracy they saw fit. But decades of fragmented rule and competition meant that factional violence became institutionalized. By 2014, Libya had reverted to a Hobbesian order, and several different parliaments existed with their own armies, the UN-backed government in Tripoli's being the weakest.[62] The hell of best intentions created a new humanitarian disaster that the average Western citizen cared little about—except for agitation over Libyan refugees fleeing by boat into Europe—while R2P cheerleaders' advocacy quickly faded.

Trend Lines of the Future: The Libyan Battlespace

Technology can directly shape a battlespace in the way military organizations incorporate technology in their exercise and management of violence. The level of scientific knowledge (broadly speaking) in a society can indirectly shape a battlespace and can become an embedded element of the battlespace, if it influences combatant decisionmaking. Gen. William Tecumseh Sherman in the American Civil War and Gen. Curtis LeMay in World War II pursued destructive actions against civilians. But to them their targeting logic was aimed at technologies of warfare that supported the war effort by breaking the enemy's will and dismantling industrial/economic capacity to resist.

Initially in war, technology is exogenous to the battlespace, but the moment combat begins, technology becomes endogenous to the battlespace because it is about not merely what technology the military has adapted but also the level of technological adaptation by the adversary and its society. The Libyan air campaign generated a novelty by way of a positive externality of an American military application and a scientific achievement that is taken for granted, though it has fundamentally transformed human activity in all corners of the globe.

In 1974, the US military proposed a tapestry of satellites around the globe. The GPS satellite system began in 1977 under the auspices of the US Air Force and is now managed by the US Space Force. The United States has provided GPS as a global public good since 1983, with the Ronald Reagan administration prompted into providing this service after the Soviets shot down a Korean passenger aircraft that had strayed into Soviet airspace.

On top of personal computers and high-speed internet access, which were transforming society, came mobile devices with cellular and computing capabilities, soon dubbed "smart devices." Soon every electronic device would have an Internet Protocol address and be anchored to a GPS system. Equivalent systems in Europe, China, and Russia have recently emerged as the costs of becoming a space-faring power have dramatically decreased.[63] The combination of personal computers, the internet, and smart devices would lead to the emergence of a new domain, with revolutionary implications, the trend lines of which were just coming into focus when the United States intervened in Libya.

Scientific knowledge, in its pure and applied forms, spreads far and wide in a diffused manner as more societies adopt it, and suddenly it becomes a taken-for-granted reality. The rise of smart devices and their ease and accessibility changed the way people engaged in day-to-day activities in unexpected ways and unexpected corners. For example, in failed states such as Somalia with no central government, citizens bypassed brick-and-mortar banks for mobile banking. Smugglers and traffickers cross the Sahara, with cheap handheld GPS devices, expertly avoided customs, border patrols, and bandits. In authoritarian states, the internet and associated technologies facilitated internal control, such as turning off cellphone networks to prevent social mobilization and easily tracking smart devices to arrest dissidents.[64]

Digital devices and the internet played a decisive but not a defining role in the 2011 Arab Spring, with cascading protests across countries.

There was a self-serving unanimity in the United States that the Arab Spring was a digital revolution. However, security agencies and governments in the region saw it differently. They referred to it as the *Al Jazeera* revolutions and were enraged with Qatar. Most Arab countries perceived Qatar's ruling family to be using journalism to actively undermine authoritarian rulers—except themselves. At the time, most people in the Middle East did not possess smart devices—they were a luxury—but every house had multiple TVs, with *Al Jazeera* beaming. The moment protests began in Tunisia, nonstop coverage of them made it possible for people to juxtapose the events elsewhere with grievances of their own. It became easier for them to superimpose the rage of a fellow citizen in the Middle East under oppressive and corrupt rule, and they could translate it into their own frustration.

Social media platforms were a societal undercurrent, utilized by a few, and tech savvy youth utilized these online platforms to coordinate activities with crowdsourcing apps, another novelty that few governments had anticipated as a platform for antiregime social movements. A preview of this phenomenon occurred in a different corner of the world under different circumstances, and the logic, in a different manifestation, would keep recurring, given the ubiquity of smart devices.

Smartphones that enabled internet connectivity and on-demand GPS began decisively shaping battlespaces with their mainstream introduction in 2007. At the time, the idea that one could fit a computer in the pocket and make phone calls was a novelty.[65] However, the rate at which they were adopted in societies—rich and poor—was unprecedented in terms of adaptation and integration of new technology. From mainstream introduction of smartphones in 2007—initially the Apple iPhone and subsequently Android devices—the number of smartphones sold annually had increased tenfold by 2014 (Figure 4.1).[66] The smartphone created a real-time information tool, an ability to access practically any information instantaneously, anytime, anywhere. The amount of progress made in less than a decade was built on rapidly compounding processes of adaptation, affordability, and integration. The intensity of similarly disruptive devices pales in comparison. For instance, the printing press was introduced in 1440, and its mainstream use and adoption (i.e., cheap books, newspapers, etc.) had become a global standard by 1800. In a process that took over three centuries, the printing press undermined governments and transformed societies, economies, cultures, and religions.[67] The advent of smartphones and interconnectivity has compressed centuries of monumental change into under a decade.

With cellular phones pairing GPS capabilities with free software— Google Earth, an inclinometer, and a compass app—untrained insurgents

Figure 4.1 Sales of Smartphones Since 2007

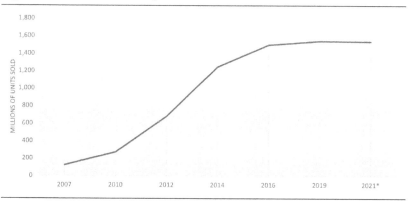

Note: *As of March 2021.
Source: S. O'Dea, "Global Smartphone Sales to End Users, 2007–2020." *Statistica.*

could quickly communicate like a regular army formation and precisely adjust and range their rocket and mortar shots. This enabled a new level of lethality and capability for nonstate combatants. This state military capability took conventional forces months to train and be proficient at, and in places like Libya and Syria, insurgents were integrating a rapidly emerging technology to bring a modicum of tactical effectiveness on the battlefield.[68] Soon NATO and the intervention entrepreneurs in the West disengaged from Libya behind eloquent self-exculpations. America was attempting a broader strategic disengagement from the Middle East, and just when the United States thought it was out, the region dragged it back in.

The Rise and Fall of ISIS

The Islamic State in Iraq and Syria (ISIS) was the territorial manifestation of the same totalizing millenarian vision that animated the jihadist terrorists who had been attacking Western targets since the 1980s.[69] The rise of ISIS was an ideational phenomenon and also a function of state failure. The Iranian Quds force worked closely with Syrian intelligence to support and maintain a logistics supply line of jihadists during the American occupation of Iraq (2003–2011). When antigovernment demonstrations turned into the full-fledged Syrian Civil War (2011–present), the Syrian state slowly began losing territorial control of Syria. In this power vacuum, ISIS emerged as a rebranded version of Iraqi jihadists working with unemployed Saddam Hussein–era military officials. ISIS exploited the smuggling networks the Syrians and Iranians had originally established to fight US troops and turned this network against Damascus. This allowed ISIS to establish territorial control across eastern Syria.

With a solidified rear base, ISIS established a series of local alliances inside Iraq with disenchanted Sunnis. Then ISIS quickly routed the Iraqi state in 2014, breaking its expensive, fragile Fabergé egg army.[70] In short time, ISIS controlled territory the size of the United Kingdom and a population the size of Jordan. ISIS implemented its vision of the ideal world: a "caliphate." The newly established Islamic proto-state levied taxes, issued passports, and even ran its own DMV.[71] Individuals refusing to subscribe to its vision were publicly executed or buried in the desert. Those who aligned their vision found in ISIS a cause worth fighting for and defending. Internationally, the vision translated into promoting the movement to a globally connected audience that wanted to watch and interact on their smart devices.[72]

The war ISIS created brought entire families from all over the Muslim world to join the ISIS movement.[73] They were not being misled; they knew exactly what they were getting into. New adherents saw in ISIS an actualization of an idea in which they strongly believed. It was possible because cyberspace had become a virtual domain of human interaction, where a cross section of people from all over the world could identify with an idea through mobile phone–enabled live streams and even interact in real time and ask questions. Recruits came voluntarily, and ISIS utilized innumerable platforms to recruit, disseminate propaganda, shape the overall discourse, and reach individuals in far corners of the world on a one-to-one basis.[74] The flexible social media approach by ISIS recruiters fused "fashion" and "spectacle" to accumulate diverse amounts of social capital in various online forums and in a range of countries.[75] The level of personalization had been unthinkable only a few years previously.

They also utilized social media platforms before, during, and after kinetic missions to shape the information environment, which is an important element of an active battlespace.[76] Social media platforms were slow to stop the use of their systems as terrorist propaganda venues, as there was neither penalty nor benefit, and because it would run counter to the monetizing logic of these firms that rely on continuous interaction. Notions of the old battlespace—namely, free market competition and that the best idea wins in the battle of ideas—failed epically. However, censorship by private companies that operated social media platforms presents the problem of nongovernment entities becoming the ultimate arbiters of truth and of good versus bad. This also presents the West with a dilemma over the value of free speech and when it crosses the line to become dangerous speech.

Knowing the level of local-global connectivity, ISIS made utilizing Western prejudices against the West its trademark. For example, the rise of transnational juridical realities meant that most nonstate actors fighting a legally constituted state learned how to turn the juridical realities into tactical advantages—for instance, by placing mortars next to children's schools, using children and pregnant women as human shields or as active combatants, and using protected spaces to tactical advantage, to name a few. ISIS formalized these attempts into a science. Simultaneously, Syrian, Iranian, and Lebanese Hezbollah forces fighting on behalf of the Syrian regime would utilize the same tactics as ISIS and develop their own sophisticated cyber tactics whereby automated bots would generate chatter on social media platforms to deflect attention before or during kinetic activity, thereby skewing the discourse in digital media. They effectively incorporated the cyber domain in its social interactive aspect to shape the information environment, aligning it with kinetic efforts. US and Western powers noticed the trend lines of technological advancements being incorporated by state and nonstate adversaries; however, they did not see them as strategic challenges, because these efforts were viewed through a tactical lens, as they emerged, and as pertaining to specific (usually) active battlespaces.

Conclusion

Several decades removed from the end of the Cold War, the West no longer retains a preponderance of power, and America can hardly claim military unipolarity. Indeed, the West in terms of its ideational foundation is slowly coming apart at the seams. While there are many contending arguments about the decline of American military hegemony specifically and Western power broadly, the fact remains that geopolitical realities have shifted fundamentally, as have ideational foundations.[77] While the West abides by transnational juridical arrangements, regional powers and revisionist powers brazenly disregard them with no repercussions—Russian, Syrian, and Iranian actions in the Syrian Civil War are examples.

Chinese actions throughout Asia, Russian actions in Ukraine, and the character of the civil war in Libya (2014–present) correlate with the shift in geopolitical and ideational realities. Nonstate combatants, if they can become proxies of a regional power or a revisionist power, also manage to disregard transnational juridical arrangements and Western concerns. In sum, the West is entering a transformed global strategic context, a transformation at variance with what was expected at the end of the Cold War and the irresponsibly managed American unipolar moment.

Notes

1. Stanton, *Violence and Restraint in Civil War.*
2. Krasner, *Sovereignty*, 84–90.
3. Thomas, *The Helsinki Effect.*
4. Havel, *The Power of the Powerless.*
5. Sharansky, *Fear No Evil.*
6. Gevorkyan, Timakova, and Kolesnikov, *First Person*, 65.
7. This qualitative edge was reflected in a key Russian request during the Strategic Arms Limitation Treaty negotiations to limit the nuclear-capable supersonic B-1 Lancer as a conventional bomber—which the United States did.
8. David Vine, "Where in the World Is the U.S. Military?" *Politico* (July/August 2015).
9. Morgenthau and Person, "The Roots of Narcissism," 342.
10. McMaster, *Battlegrounds.*
11. Fukuyama, "The End of History?," 3, 4, 9.
12. Jones and Stedman, "Civil Wars and the Post–Cold War International Order."
13. Kaldor, *New and Old Wars.*
14. Jane Perlez, "Insurgents Claiming Victory in Somalia," *New York Times*, January 28, 1991, 3.
15. Menkhaus, "Somalia," 133–159.
16. Robinson and Matisek, "Assistance to Locally Appropriate Military Forces in Southern Somalia."
17. Licklider, "The Consequences of Negotiated Settlements in Civil Wars, 1945–1993."
18. For an example of such utopian peacebuilding arguments for state-building, see Newman, "The Violence of Statebuilding in Historical Perspective."
19. Based on a 2011 estimate, these internationalized Somalian state-building costs are likely over $70 billion in 2020, given the expenses associated with the African Union Mission in Somalia and many other political and military missions by the United States, the United Nations, and various states and actors. Bruton, *Twenty Years of Collapse and Counting.*
20. Themnér, *Warlord Democrats in Africa.*
21. Jayamaha, *Rebels, Inside and Out.*
22. Olson-Lounsbery and Pearson, *Civil Wars.*
23. Bøås, "Liberia and Sierra Leone—Dead Ringers?"
24. Kaplan, *The Coming Anarchy.*
25. Jayamaha, *Rebels, Inside and Out.*
26. "Get Him to the Greek Music Video—African Child (2010)—Russell Brand Movie HD," video uploaded to YouTube by FandangoNOW Extras, December 20, 2013, www.youtube.com/watch?v=H-YCZr0epts.
27. Melvern, *Conspiracy to Murder*, 5.
28. Barma, "Peace-Building and the Predatory Political Economy of Insecurity."
29. Landau-Wells, "High Stakes and Low Bars."
30. International Commission on Intervention and Sovereignty, *The Responsibility to Protect*, IDRC Digital Library, December 2001, https://idl-bnc-idrc.dspacedirect.org/bitstream/handle/10625/18432/IDL-18432.pdf?sequence=6&isAllowed=y.
31. Anne-Marie Slaughter, "Why Libya Sceptics Were Proved Badly Wrong," *Financial Times*, August 24, 2011, www.ft.com/content/18cb7f14-ce3c-11e0-99ec-00144feabdc0.

32. Anne-Marie Slaughter, "How to Halt the Butchery in Syria," *New York Times*, February 23, 2012, www.nytimes.com/2012/02/24/opinion/how-to-halt-the-butchery-in-syria.html.
33. For example, Power, *"A Problem from Hell"*; Slaughter and Feinstein, "A Duty to Prevent."
34. "Full Committee Hearing: Nominations," US Senate Committee on Foreign Relations, March 23, 2021, www.foreign.senate.gov/hearings/nominations-032321.
35. Kaldor, *New and Old Wars*.
36. Petersen, *Western Intervention in the Balkans*.
37. Caspersen, *Contested Nationalism*.
38. The accidental airstrike—with five bombs—on the Chinese embassy in Belgrade, Yugoslavia, was not a good look for the United States.
39. The George H. W. Bush administration rejects notions that such a promise was ever made to the Soviets. However, the existence of such a rumor and its feeling of truthiness in Russia continue to resonate with leaders in Moscow. For more, see Sloan, *NATO, the European Union, and the Atlantic Community*, 154–156.
40. McKenzie and Loedel, *The Promise and Reality of European Security Cooperation*, 27–29.
41. Politi, *The Rome Statute of the International Criminal Court*.
42. Bergsmo and Buis, *Philosophical Foundations of International Criminal Law*.
43. Kilcullen, *The Accidental Guerrilla*.
44. Rubenson, *King of Kings*. As a matter of coincidence, a Tigrayan prince allied with the British and acquired their weapons and ammo in exchange for assistance on the trek out of Ethiopia. With this newfound aid, Prince Yohannes IV quickly routed opponents and crowned himself emperor, putting in place the rubrics of what later became the modern Ethiopian state. It was a propitious externality of a war for pride, since the expedition had a limited objective, much like the American-led intervention to expel Iraq from Kuwait (1990–1991).
45. Barfield, *Afghanistan*.
46. Fraser, *The Flashman Papers*. We strongly recommend reading about the exploits of Flashman serving in Afghanistan (1839–1842) in the first 1969 book published, which is a fictional account of a British military officer who is best described by the author as "a scoundrel, a liar, a cheat, a thief, a coward—and, oh yes, a toady" (11). The dark humor is that Flashman always finds a way to accidently become a war hero in every major British battle of the nineteenth century, eventually earning the rank of brigadier general. The twelve-book collection is perhaps the best and most entertaining introduction to understanding the nature of warfare during the height of the British Empire.
47. Rynning, *NATO in Afghanistan*.
48. Gall, *The Wrong Enemy*.
49. The humorous *Duffel Blog* is similar to *The Onion*. Two *Duffel Blog* articles on making progress in these wars came out in 2017 (www.duffelblog.com/2017/11/nicholson-turned-corner-afghanistan and www.duffelblog.com/2017/02/were-making-real-progress-say-last-17-commanders-in-afghanistan); both preceded a real article (Paul McLeary, "U.S. Has 'Turned the Corner' in Afghanistan, Top General Says," *Foreign Policy*, November 28, 2017, https://foreignpolicy.com/2017/11/28/u-s-has-turned-the-corner-in-afghanistan-top-general-says) in which an American general optimistically assesses the war as *Duffel Blog* had predicted.
50. Afghan ethnic demography: Pashtun (Pashto), 42 percent; Tajik, 27 percent; Uzbek, 9 percent, Hazara, 8 percent; Aimak, 4 percent; Turkmen, 3 percent; Balochi (Baluch), 2 percent; others, 5 percent. Benjamin Elisha Sawe, "The Ethnic Groups of Afghanistan," *World Atlas*, September 10, 2019.

51. Salman Masood, "Pakistan's Prime Minister Suggests Osama Bin Laden Was a Martyr," *New York Times*, June 26, 2020.
52. Butt, "Why Did the United States Invade Iraq in 2003?"
53. Today, with the benefit of hindsight, the absence of WMDs in Iraq is used as an excuse to peddle conspiratorial assertions to explain the Iraq War. What can be stated with certitude is that the US military and Special Operations Forces were exclusively tasked with discovering WMDs and spent months trying to uncover them. That suggests that WMDs were a genuine concern, even if predicated on faulty intelligence and decisionmaker biases, and that is a different task for future historians.
54. It reminded Iraqis of the British military occupation of Mesopotamia (1920–1925). For more, see Omissi, *Air Power and Colonial Control*.
55. "Bush: Join 'Coalition of Willing,'" *CNN*, November 20, 2002.
56. Burke and Matisek, "The Illogical Logic of American Entanglement in the Middle East."
57. Gibson, *Securing the State*.
58. Bogaards, "Iraq's Constitution of 2005."
59. Purity of intentions is typically filtered through a partisan lens.
60. Kawczynski, *Seeking Gaddafi*, 242.
61. Clinton, *Hard Choices*, 375.
62. Matisek, "Libya 2011."
63. Such systems synchronize for positioning, navigation, and timing, and when such a constellation of satellites works globally, it is known as global navigation satellite system. Besides the American GPS, similar systems operated include Russia's GLONASS, China's BeiDou, the European Union's Galileo, Japan's Quasi-Zenith Satellite System, and India's Regional Navigation Satellite System with plans of becoming global.
64. Brownlee, *New Media and Revolution*.
65. The introduction of the BlackBerry phone in 2002 epitomized the convergence of technology—a phone fused with a computer—as it provided rudimentary email and web browsing capabilities. Until 2007, BlackBerry phones provided basic smartphone capabilities for niche business purposes such as work emails.
66. For the most popular smartphones in 2007, see www.pcworld.com/article/140584/article.html.
67. Febvre and Martin, *The Coming of the Book*.
68. Kilcullen, *The Dragons and the Snakes*.
69. For background on contemporary radical Islam, the Muslim Brotherhood (Jamā'at al-Ikhwān al-Muslimīn), and how their doctrinaire leaders came to prominence in 1950s and 1960s Egypt, see Perego, "Clampdown and Blowback"; Qutb, *Ma'ālim fī al-ṭarīq*.
70. Matisek, "The Crisis of American Military Assistance."
71. Shadi Hamid, "What America Never Understood About ISIS," *The Atlantic*, October 31, 2019.
72. Buddhika Jayamaha, Kevin Petit, and William Reno, "Iraq's Path to State Failure," *Small Wars Journal*, April 21, 2017.
73. Reno and Matisek, "A New Era of Insurgent Recruitment."
74. Verini, *They Will Have to Die Now*.
75. Richards, "'Flexible' Capital Accumulation in Islamic State Social Media."
76. Grove, "Weapons of Mass Participation."
77. For example, Mearsheimer, *The Great Delusion*; Walt, *The Hell of Good Intentions*.

5

Compressed, Converged, and Expanded Battlespaces

THE UNIPOLAR MOMENT HAS COME TO AN END, AND INTERSTITIAL warfare has returned. Current conceptual frames are inadequate to delineate the strategic horizon where adversaries create new strategic realities using subtle maneuvers. Tyrants have reasserted themselves at the confluence of Western strategic narcissism, ideational hubris, and the incongruity between Western strategic aspirations and capabilities. Antagonists always have a vote, observing Western strategic contradictions and identifying vulnerable moments for exploitation. Knowing that any maneuver (however subtle) is bound to have strategic ramifications, many actors have devised ways of strategically acting below the threshold of a response, all without incurring the wrath of a nuclear-powered status quo power. These actions are sometimes considered gray-zone activities and in many ways circumvent traditional notions of deterrence that US and allied leaders rely on.[1] Hostile actors exploit the strategic contradictions of status quo powers by leveraging emergent domains and implementing interstitial tactics.

Status quo powers, even wielding hegemonic power, face a particular challenge—a particular burden—vis-à-vis regional powers and other revisionist and revolutionary actors. The status quo represents the way things are, reflecting a specific configuration of power relationships. A hegemonic power and its allies that benefit from the existing power configuration will attempt to maintain it, while those that do not or that believe an altered arrangement would benefit them better will always attempt to alter the existing configuration of power—whether it is ideational, economic, military, or political. Therefore, status quo powers need to be vigilant and credibly display with words and actions that they are determined to maintain the reality as is and, most crucially, will defend it by force of arms if necessary. Most importantly, they must signal the resolve and willpower to play a direct role in any alteration of the reality.

For example, the wars of the French Revolution and the Napoleonic era sought to alter the ancien régime, the postfeudal power arrangements of continental Europe based on the divine right of kings. French hegemonic ambitions failed, giving rise to what came to be known as the Congress of Vienna, a broad understanding—but not an institutional arrangement—between the sovereigns of the time: the Austrians, Prussians, Russians, British, and defeated French. The understanding of the Congress of Vienna was that they would keep France in check and maintain a semblance of order on the Continent and that Britain would intervene to prevent the rise of a hegemonic power in central Europe. This idea of balancing power worked in principle but not as well in reality, at least not as perfectly as often claimed, for if it did, then German unification would not have been possible. German unification is a defining example of altering strategic realities with the use of interstitial tactics, albeit played out in a single domain: land.

Altering Strategic Boundaries: A Warning from the Past

Prussia, the largest and strongest partner in the German confederation, engaged in series of interstitial tactics. Prussia exploited the incongruity between the certitude of principle and ambiguity of realty on the ground. Prussia took actions for small gains, exploiting the internal weakness of the Austrian Empire and Britain's legalistic and moralistic approach to foreign policy, an approach shaped by its own domestic politics. In principle, the British were committed to maintaining the status quo, the diplomatic understanding arrived at in the Congress of Vienna, which amounted to maintaining the state of affairs and collectively crushing any hegemonic ambitions. The clarity of principle did not exist in reality because the boundaries of the stated order were ambiguous and not delineated. Specifically, what agreed-upon indicators would suggest hegemonic ambition, and in what conditions would the powers commit to war in order to uphold principle?[2] The ambiguity of reality meant porous boundaries in the stated strategy. The way to find out what unclear boundaries are is to test the status quo powers, to keep pushing their logic of upholding and defending the system.

Prussia tested British commitment and stated willingness to commit to a continental war. A state can proclaim a commitment, but such a commitment only becomes a reality if that state abides by it. Thus, strategic provocateurs will conduct probing actions to find the precise contours of an openly stated commitment by a more powerful state.

The moment a provocateur can identify and isolate an incongruity between proclaimed commitments (aspiration) and the political willpower of the state to act (credibility), that incongruity becomes maneuvering space for a competitor to achieve objectives. Prussia knew that Britain had no standing army to speak of and that Britain's primary objective long had been the maintenance of maritime supremacy, which undergirded English hegemony. Britain's actions during the nineteenth century also showed that it would act unequivocally. Its people would mobilize, especially if its maritime primacy was threatened, precisely because British politicians made clear the link between naval supremacy and economic success. Similarly, a brazen challenge to its honor or any molesting of Her Majesty's subjects would also mobilize the people, compelling a military response, as the Abyssinian punitive expedition showcased.

Conceptually, the British doctrinal position was focused on containing France. With the Napoleonic Wars setting the geopolitical context, British, Austrian, and Russian leaders imagined that France would be most likely to give in to revisionist temptations, not Prussia. Louis Napoleon, the French leader who proclaimed himself emperor in 1852 and made outrageous proclamations about reinstating French grandeur, reified the sentiment among other powers that France, not Prussia, would give in to hegemonic ambitions and attempt to alter the status quo. In contemporary terms, the French leader inadvertently conducted an active misinformation campaign on behalf of Prussia that the Prussians could have neither devised nor executed better. Consequently, Prussia remained outside the scope conditions (conceptual limits) of the British strategic paradigm due to the assumption that Prussia would continue to play its role as the first among equals in the loose German federation.

In three quick wars with limited objectives—in nine short years—the Prussians built a federal and monarchist structure from Schleswig-Holstein to Bavaria under the Prussian aegis, creating a continental power that fundamentally redrew the European map. Prussia, in its German unification enterprise, outsmarted and outwitted its far more powerful peers so well that they did not realize the strategic significance of the outcome. At that point, with the strategic picture so altered and realities on the ground so well established, neighbors had no choice but to settle for Prussian expressions of goodwill, which Otto von Bismarck, the architect of the enterprise himself, delivered at the Congress of Vienna.[3] In altering the strategic picture with interstitial tactics, Prussia relied on three key tenets.

First, Prussia relied on its understanding of each adversary and its domestic political structures. Prussia observed the weakness of the Austrian emperor as the head of a polyglot empire. The Russian absolute monarchy could be decisive, but its distance advantaged Prussia. Britain had proximity but a domestic electorate to manage. Second, Prussia accurately understood its maneuvering space by recognizing disconnects between great power commitments to maintaining the status quo and the willingness to follow through on obligations. Third, land was the defining domain, and the Prussians—Gen. Helmuth von Moltke specifically—aligned the military ends, ways, and means. Moltke accounted for geography, such that all military efforts seamlessly fit into the desired political end state.[4]

Returning to the present, the contemporary logic of power that drives states' behavior in how they relate to existing international orders remains the same. However, aspirations and strategic end states of states' power manifest differently, with radical consequences for the changing character of warfare. Revisionist powers, regional powers, and assorted spoilers (state and nonstate) today increasingly exploit the altered geopolitical context. These hostile provocateurs leverage the doctrinal incongruities and contradictions of Western powers noted in the last chapter and have the capacity to leverage capabilities across the traditional domains of land, air, and sea and the emergent domains of space, cyber, and civil society in liberal democracies. That fundamentally alters the strategic dynamic, presenting expanded battlespaces with broader maneuvering space.

Interstitial Warfare Redux

Interstitial warfare took an explicit form during the Cold War in spheres of influence and in the contested sphere between the United States and the Soviet Union. In the direct spheres of each superpower—the Soviet satellite states, Soviet homeland, US homeland, North Atlantic Treaty Organization (NATO) allies, and Pacific partners—the adversaries engaged in overt and covert propaganda activities. Then as now, however, authoritarian states had an advantage against the free societies of the West. They exploited the vibrancy of free societies and the myriad ideational currents that animated people, using them against them in their own societies. The semblance of order that pervaded during the Cold War, as some see the period with a bit of nostalgia, arrived slowly over time, as both powers came to terms with each other's actions and generated a modus vivendi.

Both superpowers were explicit about their contestation and disagreements but agreed to (mostly) not disagree on their disagreements. No such understanding exists today. The unipolar interregnum is ending. Geopolitical spheres of interests, influence, and contestation are not clearly delineated, only guessed at and questioned as the transnational institutions, laws, and norms that were supposed to usher in a new era are slowly hollowed out. Thus the end of the interregnum means states, small and large, are taking active overt and covert measures to shape the emerging reality, always testing the credibility of the hegemon's proclaimed commitments. Interstitial tactics now play a defining role across the traditional domains of land, sea, and air and the emergent domains of space, cyber, civil society, and transnational institutions. Though all states wish they could alter global power realities to work in their favor, only some with global heft can turn those aspirations into reality.

Exploiting Geopolitical Realities

The realities of geopolitical power shape state, institutional, and individual behaviors. International norms, laws, and institutions are ultimately reflections of geopolitical power realities at the time they were created. Upholding the principles they espouse requires enforcement. The Rome Statute that established the International Criminal Court (ICC), for example, provides for the prosecution of those accused of genocide, crimes against humanity, war crimes, and aggression. But the ICC faces a conundrum in instances such as Russia's seizure and annexation of Crimea from Ukraine in 2014. ICC prosecutor Fatou Bensouda announced in December 2020 that her office's findings indicated that it was probable that parties to that conflict, including Russia, committed war crimes and crimes against humanity.[5] But Russia (like the United

States and China) is not a signatory to the Rome Statute, and Russia refused to cooperate with the ICC. Since Russia (like the United States and China) occupies a permanent seat on the UN Security Council, it is also highly unlikely that the United Nations will play a significant role in shaping outcomes in which Security Council members have opposing interests. This conundrum reflects the new geopolitical reality in which powerful states are much less likely to permit international institutions to play significant roles in shaping conflicts. Compare this situation to the height of US unipolar dominance in the 1990s and 2000s when armed UN "peacekeeping" operations and NATO interventions in Kosovo and Libya were met with UN Security Council permanent members' consent or at least abstention.

Some of these transnational institutions are devised in such a way that states opting into them uphold them because they have been codified into domestic legal arrangements. US domestic statute (18 US Code § 2441—War crimes), for example, directly incorporates the 1949 Geneva Convention Common Article 3, which defines and prohibits war crimes such as torture. But transnational institutions are harder to modify than domestic legal arrangements, and instead geopolitical realities frequently govern whether an international agreement can be modified, and if so, whether it will have any real force.

What transnational institutions have in common with domestic juridical realities is that one can always exploit legal loopholes or alter the legal configuration (i.e., rig the system) such that it works in a group's favor. Depending on the penalty mechanism in place, (powerful) states can pick and choose at will. These institutions are staffed by people with nationalities, despite their status as global civil servants. States can (and occasionally do) "capture" these individuals.

The end of the unipolar moment finds more instances of rhetorical appropriation of Western norms. Russia invoked the responsibility to protect to invade and annex parts of Georgia in 2008. In so doing, it laid bare the power realities that underlie such ideational enterprises. Who adjudicates when and how to intervene is ultimately an issue of power. Russian actions also probed the credibility of US and European commitments to Georgia. Russia thus called into question actual NATO and American geopolitical commitments at the edge of NATO and EU spheres of influence. Russia accurately defined its safe maneuver space and replicated its maneuvers with even more swiftness and precision in eastern Ukraine in 2014. In Georgia it employed a limited interstitial tactic with little global strategic ramification, concentrated in the land domain, in the traditional Russian sphere of interest. Similar Russian activities occur in the Baltic states but with less success because of proactive measures by each NATO state and the mobilization of civil society to be more resilient against Russian probing actions.[6]

In Ukraine, Russia leveraged the emergent domains of cyber and civil society, exploited the blind spots in Western strategic doctrine, and created an altered global strategic reality. In its annexation of Crimea, Russia leveraged the cyber domain (to create confusion and disable infrastructure) and social media (i.e., civil society) to shape perceptions. This is an important development because one finds in Russia's operations the broad elements of the old Cold War concept of political warfare that targets the ideas and perceptions of the adversary's society but applied in a manner that also directly assists operations on the ground, including in real time. One way of thinking about this development is to recall from Chapter 3 George Kennan's Cold War idea of political warfare to sow doubt in the minds of the adversary's population about the viability and legitimacy of its political system in order to achieve the strategic end of a significant shift in the geostrategic balance and even change an adversary's political regime. Set that idea alongside the contemporary American concept of military information support operations (MISO), defined as the ability to "convey selected information and indicators to foreign audiences to influence their emotions, motives, objective reasoning, and ultimately the behavior of foreign governments, organizations, groups, and individuals in a manner favorable to the originator's objectives."[7]

The post–Cold War American doctrinal approach to MISO, like most things involving information and military force, is to limit the conduct of political warfare to create effects at an operational level. This approach reflects American political culture and legal frameworks that preserve a sharp distinction between domestic and international operations. This approach is in line with the American public's expectations of firm civilian control over the US military and the exclusion of its operations from US territory or against US civilians. The Russian approach blurs this domestic/international boundary and brings Russian and Ukrainian societies in play as part of a political warfare battlespace. The role of technological developments in the GRINS framework enabled Russian operators to use information in real time to create effects at tactical and operational levels. These effects gave Russia's leaders an enhanced capacity to exploit geopolitical realities in Georgia and Ukraine to weaken the credibility and thus the power of its geopolitical adversaries.

The Great Enlargement

These are among the trend lines of an emerging operational environment that exploits US structural vulnerabilities.[8] First, there was *expansion*, as an enlargement of the battlespace in terms of time (phases), domains, geography (space and depth), and actors. Second, *convergence* permitted the astute use of technology to align military and political end states. Finally, *compression* enabled the ability to expand the battlespace and converge capabilities, which rapidly and potently compressed the strategic, operational, and tactical levels of war.[9]

Russia premised its interstitial engagement by leveraging axiomatic Western doctrinal and juridical premises to outsmart and outmaneuver the West—that is, Russia conceptually outflanked the West first. The West, since the end of the Cold War, has taken an increasingly legalistic and moralistic approach to identifying aggression. Such a sentiment became a canonical premise during the unipolar interregnum. Consequently, officials in the West and the broader international juridical and human rights community see aggression only in instances when women and men with guns engage in killing. This is a conceptual blind spot, with strategic and tactical ramifications.

When doctrinal and juridical frames fail, they become redundant conceptual notions if people with guns do not fit neatly into a certain category. This contributes to an ambiguous category of civilians, nonstate combatants, and/or whether someone can be legally framed as a combatant. Geopolitically, Crimea and eastern Ukraine (specifically Donbass) also fit within the interstitial space where Western influence is at its lowest. How willing is the West to act, in this periphery, against unidentifiable combatants (even though the intelligence community could tell you they were Russian special forces, it could not advertise this because of the collection methods employed).[10] Russia employed a successful interstitial tactic, subversively implemented, that relied on multiple domains.

Following the emergence of "little green men" in Ukraine, Russia organized Crimean civil society, volunteering residents to join Russia by way of Western norms, and locked in the veneer of legal annexation via electoral legitimacy by calling a plebiscite.[11] For the first time since the end of World War II, a state utilized violence as a means of altering sovereign borders by acquiring territory. This incident also laid bare the vacuous nature of Western guarantees to Ukraine at the end of the Cold War—the Western security guarantees were a necessary condition for Ukraine to sacrifice its Soviet-era nuclear weapons. By creating a secessionist conflict in eastern Ukraine through proxies—with adept surgical use of information warfare—Russia effectively dismembered Ukraine. It must be stated, however, that while these acts have immediate geopolitical ramifications, Russian actions have done more to turn Ukrainians writ large against Russia and toward the West. Therefore, the long-term strategic viability of this Russian gambit—no matter how well executed—remains to be seen.

Russian use of legal concepts in these conflicts points to a further important development—lawfare—which in this context is the use of legal definitions and frameworks alongside traditional military means to achieve operational and strategic ends. This use of the concept appears in a Chinese military strategy essay titled *Unrestricted Warfare*, written in 1999 by two officers of the People's Liberation Army. Their emphasis was on how China could influence existing legalistic international institutions in place of military force to achieve strategic ends.[12]

Charles Dunlap, the US Air Force's deputy judge advocate general before he retired in 2010, first used the term *lawfare* in a 2001 paper in which he reflected on the 1999 Kosovo campaign. Teams of lawyers, he noted, steered debates about the legality of the war and influenced operational planning. Dunlap was concerned that these debates were imposing a significant restriction on the effective use of force to achieve political ends. He defined lawfare as the "use of law as a weapon of war," which was "the newest feature of 21st century combat." He identified the problem of relatively weak US adversaries resorting to the strategic use of legal principles to "handcuff the United States," and like the Chinese officers, he saw that an adversary can "exploit our values to defeat us."[13] "In fact, most U.S. adversaries," he wrote, "actually see our political culture's respect for the rule of law as a 'center of gravity' to be exploited."[14]

Lawfare is just one dimension of the weaponization of information. Efforts to manipulate how observers receive and interpret information have been a regular feature of warfare. Abraham Lincoln, for example, recognized that the Emancipation Proclamation was an important tool for convincing influential antislavery campaigners among the British public that their country should not intervene on behalf of the secessionist states during the American Civil War. This capacity to frame the conflict as a war to end slavery had important consequences for US operations that were then able to blockade southern shipments of cotton to British textile mills without interference from the British navy. This is an example of the use of a (potential) adversary's values. The British public was part of that American Civil War battlespace, but not in the contemporary sense of the immediacy of information and the capacity of adversaries to focus its transmission on specific segments of society in real time, an element of the battlespace's enlargement dealt with in more detail in the next chapter.

Syrian Civil War

In the Syrian Civil War, Russia and Iran defined their limits of political and military maneuver space by recognizing Western contradictions—and exploiting them. The Russians and Iranians proved their critics right. Their military actions involved massive violations of civilians' basic human rights yet effectively managed to win a war in defense of the Assad regime in Damascus. It was a war thought unwinnable given its complexity and innumerable participants, which included varying degrees of militaries, insurgents, militias, and proxies, as well as a host of other actors and spoilers.[15]

In Syria, the Russians (and Iranians to a degree) perfected the use of social media platforms by aligning social media manipulations with kinetic missions on the ground to shift narratives, either to direct or to deflect attention. Moreover, they employed various techniques and capabilities across the electromagnetic spectrum, jamming and intercepting various types of signals to their benefit. Disregard for international laws and humanitarian norms by Russia and Iran also displayed the logic of tactical appropriateness, not these ideational concerns. This paid immense tactical dividends in the fight against anti-Assad rebels, allowing them to outflank and outfight the rebels, who used civilians as shields.

The utopian construction of idealized geopolitical realities that developed in the shadow of US unipolar dominance kept coming into conflict with realities on the ground. The crucial test in this conflict concerned whether outright violations of these ideals could occur in ways that avoided the meaningful impacts of sanctions and contributed to actors' military means and political ends. The Islamic State in Iraq and Syria (ISIS), for example, adopted an explicit tactic of using civilians, including children, as human shields. In doing so they provided the rationale that norms of civilian protection inhibited Western and Western-backed forces from military action against ISIS in these situations. ISIS fighters converted schools and other "protected status" buildings into defensive fighting positions, with children and other noncombatants inside. US and allied attacks on these positions would constitute war crimes. Avoiding this would require sending soldiers into a well-fortified fighting position to engage in close-quarter battle, in effect sacrificing them for the sake of good intentions.[16]

Russians and Iranians could disregard Western criticism because they knew they had preemptively leapfrogged any possibility of strategic containment. They overwhelmed media outlets with varying accounts of the war in Syria to introduce just enough skepticism about the reliability of every news report and narrative. Meanwhile, US and European negotiators focused on securing Iranian agreement with the Joint Comprehensive Plan of Action (JCPOA) to limit Iran's development of nuclear material and its robust ballistic missile program. The Iran approach left them relatively free to consolidate regional influence in Lebanon, Syria, and Iraq.

South China Sea

Chinese behavior in the South China Sea utilized similar interstitial tactics, albeit using maritime militias—soldiers disguised as fishermen as a part of an irregular naval warfare strategy as engineers began construction of islands atop atolls that were defined in international legal terms as part of the marine environment beyond any state's territorial claim.[17] In doing so, China outflanked existing doctrinal frames as it established realities on the (newly constructed) ground. Meanwhile, Chinese officials calculated that US forces would not move aggressively against unarmed fishing vessels, even if it was clear that they were an advance guard for the island construction crews and that dislodging them, once they were in place, would require the use of real force against a nuclear-armed adversary.

China's construction of artificial islands to base its military forces in the South China Sea was a direct challenge to the foundational US doctrine of maintaining the freedom of navigation on all the world's oceans. That doctrine is a critical part of the US strategic interest in an open global economy supportive of US power. This principle of freedom of navigation is globally recognized in the UN Convention on the Law of the Sea, which the artificial islands showed was not enforceable in this instance. The Chinese move also created doubts in neighboring states about US commitments as an alternative to the expanding Chinese sphere of influence, particularly amid the Donald Trump administration's record of weak support and inconsistent and amateurish statements of American priorities.[18]

Simultaneously, China has used its participation in multilateral trading regimes to its advantage by rigging the system to its advantage, becoming an economy on par with the United States in under two decades. China's admission to the World Trade Organization in 2001 was intended to further integrate China into the US-dominated global economy. China has benefited greatly from participation in the global economy, a development that distinguishes its current status from the Soviet Union's relatively marginal position in global economic affairs during the Cold War.

This economic interdependence highlights within the GRINS framework how the specific geopolitical context (rising political tensions, continued economic interdependence) and regime type (the "China model" of the authoritarian state's subsidies to promote strategic industries, the cultivation of economic dependence among trading partners, forced technology transfers, and economic espionage) shape the character of the emerging battlespace. In a good example of coercive use of trade dependency against a longtime US ally, in late 2020 China imposed punitive trade sanctions against Australian products—China buys about 40 percent of all of Australia's exports—after that country's government considered conducting its own investigation into the origins of the Covid-19 virus and formally rejected China's claims in the South China Sea, aligning itself more closely with the United States.

The punitive action against Australia is one instance in which China has developed a distinctive "weaponization of interdependence" that complicates American efforts to coordinate with countries closer to China to block the further expansion of China's sphere of influence.[19] Meanwhile, the Trump administration's use of Section 232 trade sanctions—legislation enacted in 1962 as a tool to block imports that threaten national security—against Canada and Mexico impacted its relations with other allies and potential allies in the South China Sea region. Officials in these countries had to consider whether the United States would support them against Chinese pressure while the White House damaged its long-term relations with important partners for short-term domestic political gain.[20]

Middle East

The enlargement of the environment of contention to exploit US vulnerabilities was underway in the Middle East too. The US invasion of Iraq cleared the way for Iraqi exile groups beholden to Iran to return to Iraq. Some of their leaders became de facto Iranian proxies inside the Iraqi parliament. Then the US push to defeat ISIS created a military end state that was aligned with the political end states of Iran and Syria. Finally, US and European negotiations produced the JCPOA to limit Iran's nuclear program. The agreement's provisions included relaxation of harsh economic sanctions against Iran's government and the return of some of the assets seized after the shah's downfall in 1979. The agreement also opened a path for foreign investment in Iran's struggling petroleum industry and lifted many sanctions that prevented companies from doing business with Iranian enterprises. In sum, the agreement provided an economic lifeline to the Iranian regime as it solidified its hegemonic position in the Middle East.

Iranian leaders knew the deal would advance the economy and generate some level of performance legitimacy for the regime since the Islamist theocracy has little ideological legitimacy in the eyes of the Iranians. Iranian leaders also knew there would be no real consequences for violating the deal once they had enough fissile material to make their first nuclear weapon. Iranian negotiators insisted on a sunset clause in the JCPOA, which the Americans granted.

The Trump administration pulled the United States out of the agreement in 2018 and attempted to set in motion a policy of containment.[21] The Iranian objective is to leapfrog containment and use interstitial tactics to do so. Iran continued its probing actions in the air, sea, and land domains repeatedly, and these reached a new level of aggressiveness under the JCPOA. This new tactic was "reconnaissance by fire" to delineate maneuvering space while incrementally expanding it. For instance, Iran built a sophisticated air-defense system for the Houthi rebels in Yemen, while denying that the Houthis were Iranian proxies. Then Iran provided the Houthi rebels with missiles and assisted them with sophisticated drone attacks against Saudi Arabia. Iranian Revolutionary Guard Corps (IRGC) naval vessels began aggressive maneuvers around US vessels in the Arabian Gulf, locking radar on US aircraft flying in international airspace, and shot down a US Navy RQ-4A Global Hawk drone in 2019. In each instance, the United States refrained from kinetic action and just increased sanctions and rhetoric. Then Iran conducted a daring drone attack on Saudi Arabia, via Iraqi militia proxies, nearly crippling its oil industry for a month. The United States still refused to act. The Houthis attacked Riyadh with missiles, and the United States did not act. Interpreting American inaction as continued disengagement from the Middle East, Iran attempted what amounted to a coup d'état inside Iraq.

Meanwhile, Iraq's government struggled to bring Shia militias under government control, a goal that would undermine Iranian influence inside Iraq. Iran quickly used its proxies to frustrate this effort to centralize government control over all security forces and attempted, with the use of its proxies in Iraq's legislature, to create an autonomous status for Shia militias.[22] This effort reflected Iranian practice elsewhere in the region and in this case would have meant a powerful Iraqi military that is beholden to Iran, similar to Iran's backing for Hezbollah in Lebanon. To pull off the dénouement, with ISIS almost defeated, Iran lobbied Iraqi officials through proxies to get the Americans out of Iraq. Iranian-backed Shia militias engaged in tit-for-tat actions against US troops, with the Iranian Quds force commander and supreme leader using social media to make explicit threats against US troops in Iraq. All the while, the social media operation reiterated Iran's position as a victim, thereby playing into pro-Iranian and anti-American sentiments. For instance, the IRGC commander openly stated, "You [United States] will start this war but we will be the ones to impose its end. Therefore, you have to be careful about insulting the Iranian people and the president of our republic. . . . You know our power in the region and our capabilities in asymmetric war. We will act and we will win."[23] Iranian messaging was a signal to the Americans and an attempt to deter Iraqi officials and personnel from working with the United States and its allies across the sectarian divide. Many Iraqi officials, however, wanted the United States to maintain a troop presence as a hedge and counter against Iranian influence.

The United States had to make an unequivocal point by redrawing strategic contours that could not be crossed without costs. The Trump administration then ordered the assassination of the Quds force commander, whom everyone thought was invincible. As the shock of the death subsided, Iran reverted to its usual low-intensity interstitial tactics of shadowing American vessels, monitoring airspace (without locking radar), and trying to trick American aircraft into flying into its airspace. Inside Iraq, Iran contends with groups that question its agenda. Many Iraqis resent their country's being turned into an economic, political, and military client-state of Iran.[24]

The killing of the Iranian Quds force commander also reflects the geopolitical reality that the United States and European Union created but that adversaries use against them. The targeted killing of this general generated a narrative in the West focused on assessing the legality and morality, not the strategic necessity, of the act with almost no discussion of linkage to the actions of Iran and the IRGC and the wider strategic environment in the Middle East. The prominence of these concerns reflected the Cold War–era victory, with moralistic and legalistic frames that inform recent Western military doctrine. However, this legal-moral lens over the last three decades has left a conceptual blind spot. Instead of interpreting events in relation to military and political objectives, the legal-moral lens provides an opening for adversaries to shape and frame their acts to claim that these are legitimate under the West's own rules, much as China attempts to do with its claims in the South China Sea. Then, when Western military forces act in the broader pursuit of political ends (such as reacting to Iran's efforts to dominate other countries in the Middle East), Western and especially US uses of military force draw criticism from their electorates and the global humanitarian community. The Western conundrum is that, although these groups and their criticism reflect valued domestic and global ideals, their criticism is liable to become one of the interstitial tools that Iran is able to turn against US efforts to respond in this interstitial space.

If interstitial tactics involve discrete levels of violence in traditional domains, then interstitial tactics in emergent domains are difficult to interpret because it is problematic to perceive violence in intangible domains. Thus, while the domains of space, cyber, and civil society are increasingly discussed, they remain misinterpreted due to the absence of tangible violence.

Interstitial Tactics and Emergent Domains

Outer space, cyberspace, and civil society are inextricably linked. At the same time, they each have their own distinctive elements as media for larger processes that are transparent to most. These domains are much like the water from the allegory of the old goldfish saying to two young goldfish, "Morning, boys, how's the water?" The young goldfish look at one another puzzled, and one asks, "What the hell is water?" As noted by David Foster Wallace, "The point of the fish story is merely that the most obvious, important realities are often the ones that are hardest to see and talk about."[25]

Outer space has become an emergent domain by virtue of human ingenuity. Advances in technology mean that space is no longer the exclusive realm of great and powerful states, as even Nigeria has become a space-faring nation and Kenya advertises a national space strategy despite not having any assets in outer space.[26] Space has become a public and private domain that is congested, contested, and competitive.

Space is *congested* because there are limited spots around Earth to place satellites in an economically viable manner for both states and private enterprises. It is *competitive* because states and private entities compete to gain, advance, and maintain their edge. It is *contested* because there is no international consensus about where the air domain becomes the space domain.[27] There is no adjudication mechanism for how to administer behavior in space, except for the certainty that every individual—from the nomad in the Sahel with a smartphone to the city dweller with smart devices in his kitchen—is anchored in space in terms of location, time, and navigation. Space has become a highly competitive domain without which day-to-day activities in most of the developed world would likely stop.

The United States is no longer the sole hegemon in the heavens, as a growing number of countries are beginning to provide positioning, navigation, and timing as a global public good. Numerous countries now have fully functional GPS systems with global coverage, with China and Russia actively pushing the adoption of their systems by private entities and states. These are strategic plays with the long game in mind. The play fits into their respective mechanisms of domestic control whereby they will be able to control their own authoritarian version of the internet. It contributes to the fragmentation of the internet into *the internets*, relying on varied governance mechanisms. To create a controllable internet, tied to the authoritarian control of the state, first requires that it be anchored to a GPS system, which China and Russia now have; this will be followed by imposing Westphalian sovereign rights on the internet. China, as the largest holder of virtual real estate in cyberspace, now has created a sovereign border with a Great Wall around the Chinese internet.

Governments and private organizations have launched satellites into space without permission from anyone. Multiple states have deployed and tested antisatellite weapons systems, creating debris fields, endangering the International Space Station. Other than the reality that space will amplify dependencies and become more linked to the way we live on earth, it will remain an infinite frontier with no (enforceable) rules for the foreseeable future. This also means taking seriously the possibility of a Kessler syndrome scenario where a debris field in low-Earth orbit (LEO) has a cascading effect in destroying other objects in orbit until a critical mass point is reached, which destroys all satellites in LEO.[28] "Interstitial tactics" (in quotes because there are no demarcations in infinite space, rendering "interstitial" meaningless, but tactics matter) in space simply means brazen attempts by China and Russia to move satellites near Western communication satellites, and so forth, and accosting satellites that are interpreted as dry runs in developing antisatellite weapons and tactics. Space is also a domain where it is near to impossible to hide satellites, as many researchers publicly engage in tracking of all objects orbiting Earth.[29]

Moreover, private companies, such as Rhombus Power Inc., sell software licenses to Guardian—a hybrid cloud artificial intelligence service for capturing all domains of data—giving anyone with enough money the ability to identify satellite maneuvers and other vehicles' movements and behaviors in all domains.

The greatest danger with GPS is not how GPS-guided munitions deliver their precise explosives. Rather it is the breadth and depth of GPS interdependencies of a modern functioning state in the developing world, to include transportation, financial, agricultural, and military dependencies. While GPS has enabled tremendous capacities in these sectors, these increasingly expose critical vulnerabilities with technological advances and techniques to jam, spoof, and create phantom GPS signals. There are several well-documented instances of China and Russia employing "smart jamming," with one of the strongest false GPS signals emanating from Russian controlled Khmeimim Air Base in Syria.[30] This suggests that electronic warfare, if properly employed, would nullify the capabilities of weapons platforms and systems that rely on GPS for basic functioning. Moreover, software companies such as Babel Street sell data-tracking information to governments and police, precisely because every smart device—with a specific Internet Protocol address—leaves a breadcrumb trail of behavior via GPS, all through openly sourced and available public data. This can easily translate into a hostile government using this data to track the activities of individuals and their patterns of behavior, contributing to an authoritarian government understanding a complex network of individuals that espouse antiregime views. Worse, this sort of data opens numerous avenues for adversarial states to weaponize this information in pursuit of certain strategies.

Conclusion

Such is the extent of the expansion of the battlespace that the character of warfare has undergone considerable change. Yet the nature of warfare remains constant. As in the past, dominant powers face asymmetric challenges from rising powers and weaker adversaries that seek to exploit the very practices and resources that are the basis of the dominant actors' strength. These interstitial spaces have undergone the "Great Enlargement," as this chapter has shown. The next chapter turns to a different dimension of this Great Enlargement: how technology and social change have asymmetrically exposed the domestic societies of even the most powerful countries to challenge from dedicated adversaries.

Notes

The authors are expanding on the operational level multi-domain concepts from this document, placing it in the broader strategic context, https://www.tradoc.army.mil/wp-content/uploads/2020/10/MDB_Evolutionfor21st.pdf.
 1. Matisek, "Shades of Gray Deterrence."
 2. Kissinger, "The Congress of Vienna."
 3. Breuilly, *Austria, Prussia and the Making of Germany*; Dwyer, *Modern Prussian History*.
 4. Zuber, *Inventing the Schlieffen Plan*, 120–121.
 5. Iryna Marchuk and Aloka Wanigasuriya, "The ICC Concludes Its Preliminary Examination in Crimea and Donbas: What's Next for the Situation in Ukraine?" *EJIL:Talk!* December 16, 2020, www.ejiltalk.org/the-icc-concludes-its-preliminary-examination-in-crimea-and-donbas-whats-next-for-the-situation-in-ukraine.
 6. Veebel and Ploom, "Are the Baltic States and NATO on the Right Path in Deterring Russia in the Baltic?"
 7. Joint Publication 3-05, *Special Operations* (Washington, DC: Joint Chiefs of Staff, 2014), xii.
 8. The US Army, in realizing the emergent domains, came out with an assessment of the future operating environments as "compressed, converged, and expanded." Our discussion isolates some of the shortcomings by building on their findings. Gen. Stephen J. Townsend, *The U.S. Army in Multi-domain Operations: 2028* (Fort Eustis, VA: TRADOC, December 2018).
 9. Gen. David G. Perkins, *Multi-domain Battle: The Evolution of Combined Arms for the 21st Century* (Fort Eustis, VA: TRADOC, December 2017).
 10. Galeotti, "Hybrid, Ambiguous, and Non-linear?"
 11. Mölder and Sazonov, "Information Warfare as the Hobbesian Concept of Modern Times."
 12. Qiao and Wang, *Unrestricted Warfare*.
 13. Col. Charles Dunlap, "Law and Military Interventions: Preserving Humanitarian Values in 21st Conflicts," Peter Feaver's Home Page, https://people.duke.edu/~pfeaver/dunlap.pdf.
 14. Dunlap, "Lawfare 101," 11; https://scholarship.law.duke.edu/cgi/viewcontent.cgi?article=6434&context=faculty_scholarship.
 15. Kofman and Rojansky, "What Kind of Victory for Russia in Syria?"
 16. The Battle of Mosul (November 2016–July 2017) epitomized the need for Iraqi forces to destroy most of the city just to save it from ISIS, as best shown in the 2020 documentary *Mosul*. Also see Arnold and Fiore, "Five Operational Lessons from the Battle for Mosul."
 17. J. Michael Dahm, "Beyond 'Conventional Wisdom': Evaluating the PLA's South China Sea Bases in Operational Context," *War on the Rocks*, March 17, 2020.
 18. Such as in this document that outgoing Trump officials declassified in early January 2021: "U.S. Strategic Framework for the Indo-Pacific," Trump White House Archives, https://trumpwhitehouse.archives.gov/wp-content/uploads/2021/01/IPS-Final-Declass.pdf.
 19. Farrell and Newman, "Weaponized Interdependence."
 20. Congressional Research Service, "Section 232 of the Trade Expansion Act of 1962," Federation of American Scientists, December 9, 2020, https://fas.org/sgp/crs/misc/IF10667.pdf.
 21. The JCPOA, referred to as the "Iran nuclear deal," was implemented by the Barack Obama administration on January 16, 2016, and was revoked by the Trump administration on May 8, 2018.

22. Knights, "Soleimani Is Dead."
23. "Iran General Threatens Trump: 'If You Begin War, We Will End It,'" *Al Jazeera*, July 26, 2018.
24. Author fieldwork in Iraq, 2018–2021.
25. "THIS IS WATER! by David Foster Wallace," video uploaded to YouTube by Char Mosley, December 15, 2013, www.youtube.com/watch?v=eC7xzavzEKY.
26. Eleven countries are capable of launching objects into orbit, and over eighty countries claim to be space-faring; see OECD Space Forum (www.oecd.org/innovation/inno/space-forum). Kenya even has a National Space Secretariat in charge of Kenyan space policy; see www.mod.go.ke/?p=1932. Kenya also has the Space Generation Advisory Council (SGAC); see https://spacegeneration.org/regions/africa/kenya.
27. For instance, there is the Von Kármán line, which is at an altitude of 100 kilometers (62 miles) above sea level and is typically used in treaties. However, the US military and NASA define outer space as starting at 50 miles above sea level. Finally, there is even a vague orbiting line that defines outer space as somewhere between 100 to 160 kilometers above sea level.
28. Johnson-Freese, *Space Warfare in the 21st Century*, 18–19.
29. PREDICT (www.qsl.net/kd2bd/predict.html) is one such open-source website that provides free software and tools for tracking and identifying satellites in LEO. Lily Hay Newman, "Hackers Are Building an Army of Cheap Satellite Trackers," *Wired*, August 4, 2020.
30. Greg Milner, "How Vulnerable Is G.P.S.?" *New Yorker*, August 6, 2020.

6
Civil Society and the Contemporary Battlespace

ADVERSARIES HAVE FOUND NEW MEANS TO TURN THE OPEN SOCIeties of liberal democracies into novel battlespaces. A fundamental characteristic of the open societies in these democracies is that individuals have the freedom to seek information according to their own wants and needs, to express their views and critique the views of others, and to form their own associations free of government control. As seen in Chapter 3, US officials were concerned that the Soviet Union could exploit this openness to engage in political warfare. This fear of Communist subversion led to investigations and bodies such as the House Un-American Activities Committee, first authorized as a special committee in 1938. The trial and conviction of Julius and Ethel Rosenberg for atomic espionage in 1951 fueled fears that Communist subversion endangered US national interests. Though accusations of subversion were often off the mark, the declassified report of the US government's Venona Project counterintelligence operation and its successors between 1943 and 1980 identified a persistent Soviet capacity to recruit small numbers of actual Soviet agents in the United States and its allies.[1]

The small scale of probable Cold War–era Soviet subversion reflected the GRINS reality of the time. Limited communications technologies and severely constricted flows of people and information across the Iron Curtain dividing the two adversaries' spheres of influence insulated each side from widespread subversion. The necessity of personal contact to identify and sustain individuals-turned-agents made this enterprise risky and labor-intensive for all involved. Potential agents of influence such as journalists operated in an environment with few media outlets. The presentation of news was easier to monitor and fell reliably within the spectrum of elite narratives about the United States and its place in the world. Broadcasters and newspapers contributed to cohesive national narratives through what was essentially a news monoculture. In practical terms, there was not much news to follow. Take one midsize town: television in Rawlins, Wyoming, prior to the 1981 arrival of MTV and CNN via cable service was essentially limited to Casper-based KTWO via a signal translator. Listeners to ABC News Radio heard "Hello, Americans! This is Paul Harvey! Stand byyy for Newwws!"

Contemporary residents of Rawlins or anywhere else with a smartphone or other internet-linked device, which now means everywhere, receive information with ease from a vast array of media that can be located anywhere across the globe. Pre-internet broadcasters, particularly those associated with national networks, tended to factcheck, but then as now there was no legal compulsion in the United States to verify statements, short of avoiding gross libel.[2] The old elite consensus on truth no longer exists due to the way media outlets monetize information, and some use outrageous and provocative words and images as "click bait" to increase page visits to generate advertising revenue. In that revenue model, objective truths matter less, and subjective truths become more desirable as they often are more profitable than actual facts. It is within this broader context—the rapid social and technological shift in the GRINS framework that gathered speed in the unipolar moment—that Western civil societies now are leveraged into a new battlespace.

The Virtual Selves Within Cyberspace

Cyberspace is a paradoxical construct. It is a structure where the constitutive parts, individuals, devices, and so forth, generate a virtual reality that in its totality constitutes the cyber domain. This virtual reality is an inextricable part of most people's everyday experience. Cyberspace also constitutes the sinews—something similar to a central nervous system—that ties individuals to their devices, thermostats, refrigerators, vehicles, credit cards, ticket counters, and so on, in the broader virtual reality. It generates an infosphere—an information environment that is also a datasphere—that is constantly adapting and evolving in response to imperceptible changes and societal currents, domestically and internationally.[3]

This duality of cyberspace makes it possible for individuals to inhabit the virtual reality and the sentient reality simultaneously. As individuals, organizations, and physical objects get ever more integrated and devices converse with each other without the knowledge of their owners, individuals, entities, and devices create a shadow virtual reality that is nearly identical to and parallels sentient reality. In this virtual reality, individuals exist as virtual avatars, as data reflections of themselves. Advances in storage, computing, and rapid connection speeds mean private entities monetize the virtualized selves of individuals. In doing so, private entities have created an environment of impunity, with no responsibility or accountability, other than a terms-of-service agreement that no user reads, let alone declines, since so many modern conveniences have been baited through acceptance of surrendering data to providers.[4]

It is almost as if modern citizenry in the West is dependent upon entering into the implicit contract that to exist and function in society requires one to no longer have data privacy. Efforts to reverse this development exist, such as the European Union's General Data Protection Regulation (GDPR), which went into effect in 2018. This regulation is designed to give individuals (what GDPR calls "data subjects") more control over personal data, primarily by forcing enterprises to design data protection in the development of business processes for products and services.

GDPR notwithstanding, companies monetize individuals by turning virtual avatars into real-time surveillance sentinels of their human counterparts as long as they are connected. Many industries now require massive quantities of data to operate efficiently and competitively. GDPR has created tensions within the European Union, as American companies that monetize data more readily gain commercial advantages over European competitors. In today's terms, data collection is continuous as smart devices have become essential to daily life. We know they have become essential in modern daily functioning because 95 percent of Americans own a cellular device, and there were over five billion mobile users in the world as of 2019.[5] While connectivity contributes to improving human living standards, it becomes a critical vulnerability that adversaries can and do exploit.

Cyberspace as a domain and complex network emerged in the West during the unipolar moment when US events and democracy, free market capitalism, globalization, transnational juridical realities, and other such ideas were believed to be transforming the world. As a result, the emergence of cyberspace was shaped by Western ideas, norms, and laws. This virtual reality was to be an extra layer locking in Western gains, making it inevitable that tyrants would fall, democracies would thrive, and the free flow of information in this borderless virtual reality would bring people together. Thomas Friedman's 2005 best-selling *The World Is Flat: A Brief History of the Twenty-First Century* captured the sense of that time in his idea of the world converging on a level playing field of commerce in which new technologies and multinational corporations would help unleash the creative potentials of individuals. Though not exactly reflecting the "strategic narcissism" of the American foreign policy elite—Friedman saw the United States as needing to rush to stay ahead in this globalizing world—his idea shared the strategic narcissists' confidence that post–Cold War developments would put an end to old-fashioned geopolitics.[6]

New commercial opportunities that cyberspace afforded on a truly global scale led to a great deal of business-sector faith in this virtual reality. Prescient innovators perceived how virtual reality would parallel sentient reality and how the ability to control the virtual reality would allow one to shape the sentient reality. Many of these entrepreneurs were American and relied on business-friendly US legal structures and continued US government funding for consortiums of universities and businesses to engage in innovative research. The National Science Foundation's NSFNET, for example, in the mid-1980s underwrote the networking of research university computing resources and by the early 1990s supported efforts of academic researchers and commercial users to consolidate a self-governing and commercially viable internet.

Various federal laws to protect the privacy of data in the United States are primarily focused on the protection of individuals' information (such as the Health Insurance Portability and Accountability Act of 1996 and the 1974 Family Educational Rights and Privacy Act). A plethora of state laws regulate data protection, but there is no federal requirement that companies must protect individuals' data on the scale of the European Union's GDPR measures. Moreover, US companies have a legal right to export the data of people living in the United States to countries where that data could pose a national security risk, though there are recent federal legislative efforts to limit data transfers. This (recent) patchwork of regulations offers only partial protection from efforts of foreign governments to gain access to smart device data (i.e., GPS tracking info, Internet Protocol addresses) of Americans, including lawmakers and senior military officials, and does not prevent illicit collection of such data.

Data collection of this sort, including from hacking and other illicit breeches of protections, falls under the category of espionage. In this respect, the technological and commercial aspects of this cyberspace greatly simplify an adversary's task of collecting large amounts of detailed confidential information about people and organizational capabilities. Regulators are aware of this problem. Though challenging to address, the emerging US framework appears to be moving toward GDPR-style protections. The overall effect is to create Internet blocs, with a US-EU regime of data management that increasingly departs from Russian and Chinese standards. Were an early-twentieth-century expert in what we now call data management to somehow use a time machine to come to the present, that person might find the technology strange but probably would readily recognize the problem as one of *espionage*, the use of spies to obtain politically and militarily useful information.

The more significant development, however, is in the vastly increased technical capacity of foreign actors, including state intelligence agencies, to make direct contact with individuals in other countries. This is a significant shift in the character of political warfare, as it eliminates the physical and logistical barriers of Cold War era *subversion*, the undermining of the power and authority of existing institutions and political systems.

The Undefended Civil Society

The ideational bias inherent in liberal political culture—in the classical sense that highlights individual freedoms—is that individuals have a fundamental right to free access to information. The underlying assumption is that free access to information will promote a critical mode of thinking. This kind of political culture is what Cold War institutions like Voice of America and the United States Information Agency were supposed to promote. The collapse of Communist regimes at the end of the Cold War greatly contributed to the spread of open societies, the affirmation that individualism, democracy, and reason would prevail. Popular commentors such as Francis Fukuyama and Thomas Friedman were confident of the strength of open societies and their liberal political cultures. The connectivity of the Internet, bringing vast amounts of information to the individual, was supposed to reinforce and solidify this shift to open societies. US President Bill Clinton in 2000 wished Chinese authorities "good luck" in their efforts to control online speech because "that's sort of like trying to nail jello to the wall."[7]

Faced with the juggernaut of an unavoidable virtual reality that can mirror sentient reality, the character of information and assumptions about the relation of information and critical thought have come into question.[8] This reality is a sociopolitical-information environment that is rapidly being shaped by domestic and international events, with "influencers" and "bots" able to drive debates and generate narratives in social media in a way that does not reflect reality for those not plugged into social media. This contributes to competitive ideologies of objectivity based on shared values and sociocultural interpretations, which means social engineering can take place from the bottom up, from the top down, and laterally. The competition of ideas was an element of Cold War contention that took much more of a top-down dimension, commensurate with the technology and organizational features of the GRINS framework of that time.

Contemporary internet-based companies build global dream palaces that permit actors to feed information from many directions and without the context or social space for reflection that one enjoyed during Paul Harvey's monologue on the ABC News Network, listening to his views on the national debt, welfare cheats, and bureaucrats who made no sense between his pitches for Comfort Rest mattresses and Hi-Health dietary supplements. The contemporary reality is a cacophony of social media–based rumors feeding into sectarian narratives of conflict. Or worse, the state can capture social media companies, like the Myanmar government did with Facebook, stoking and pushing anti-Islamic narratives that led to genocidal acts against the Rohingya, killing approximately 24,000 in 2016 and 2017.[9]

Governments find it much easier to regulate broadcast media and newspapers. It is more difficult to regulate social media content. The American approach to social media (as with other media) revolves around the First Amendment principle that individuals have a right to free expression in a "marketplace of ideas" and government should not infringe on that right (though in American legal doctrine, social media companies have a right to "de-platform" users whose speech violates company standards). Indeed, since Twitter's blocking of Donald Trump's account, the emphasis in American debates about social media and rights of expression has centered on whether companies should be compelled to provide unfettered access to users.

The social media–state relationship is different in China. Bill Clinton's quip above notwithstanding, Chinese Communist Party (CCP) leaders acted to restrict the new "social space" that could empower domestic actors to mobilize against corruption and authoritarian governance.[10] Virtual reality, they understood, was a new medium for the exchange of data, facts, and commerce, which could also work to formulate ideas that might run counter to the Communist Party line. The reverse would also be true. To control and shape the virtual reality would also be to control and shape the sentient reality and how one perceives reality, as private companies already demonstrated. The CCP set a clear precedent where cyberspace would evolve with distinct limits and boundaries: it would be a sovereign internet.[11] This has meant a splintering of different realities in each country, where subjective truths are transformed into facts by those in power via tight control of the internet.

Chinese officials exploit the idea of a borderless cyberspace to benefit China's national interest as they define it. There are no borders because Western countries choose not to erect borders and walls. In a borderless world, the West stood to benefit if one subscribes to the notion that unfettered flows of information lead to open societies. As the 2011 Arab Spring and the overthrow of the governments of Tunisia and Egypt showed, free flows of information can contribute to regime change. China, like many other authoritarian states, began to build virtual ramparts, creating forms of virtual sovereign territory well before the events of 2011 in the Middle East. China immediately created a sovereign border around its cyberspace, termed the "Great Firewall" in 1997.[12] China employs millions of people with the job of controlling access to and the flow of information, the way a state controls immigration and emigration. Unsurprisingly, Beijing even employs organic content generators for regime control and legitimacy by having an average-looking Chinese citizen and bots contribute to social media discussions with pro-CCP talking points, ensuring the appearance of positive Communist narratives without their discourse looking official.[13] It is no longer the sort of pre-internet hackneyed propaganda that was too cliché for the average citizen to take seriously.[14]

Any company that generates revenue with Chinese citizens must abide by Chinese rules and regulations. This creates a novel strategic departure in a world supposedly dominated by Western rules and institutions. Companies that supposedly are the guardians and champions of privacy in the West must do the exact opposite if they want to make money in China and other authoritarian states. They must store their cloud data on Chinese servers; hence the government has access to the personal data of everyone living within Chinese territory. This is the sovereign prerogative of the state—and written into law. Companies that refuse to abide by Chinese strictures will be denied access and economic gains. China thus became the first state to ascertain and demarcate sovereignty within virtual reality. Not even the honor of naming the

Facebook founder's newborn kid, as Mark Zuckerberg offered the president of China, could sway the virtual sovereignty premise that animates China's approach to the virtual reality. Flattery could not undermine Chinese strategy and understanding of cyberspace; Facebook was denied access to China. Instead China has an alternative social media platform, WeChat, with all data collected by CCP officials to conduct massive surveillance of over one billion users.[15]

The US approach of a borderless virtual space with an ideological insistence on a flattening world leaves the virtual avatar of the same person free of laws and regulations that would limit others' capacities to access that individual online. Instead, the virtual avatar of every human being is subjected to the monetizing objectives of private enterprises and the virtual laws that govern them as they control the virtual avatars. In so doing, the masses of virtual avatars are turned into surveillance sentinels, usually without the individual in any way thinking or concerned about this prospect. As noted earlier, the virtual monetizing logic is about creating ever more interactions and virtual engagement between every member of civil society, and this logic allows adversaries to directly shape, influence, and undermine Western civil society.[16]

Digital avatars—defenseless by design—of every human being across social media platforms create a permeable, undefended attack surface. This vulnerability is directly leveraged by adversaries. They engage the defenseless virtual avatars, knowing that this will generate explicit direct outcomes for society writ large. This vulnerability can be exploited through virtual societal warfare techniques, which "manipulate or disrupt the information foundations of the effective functioning of economic and social systems."[17] Adversaries access the defenseless virtual avatars of citizens in the West and manipulate them relying on *schismogenesis* (the creation of division) to turn civil societies into a new battlespace.

Civil Society

Exploitation of civil society remains one of the greatest threats that remains underconceptualized due to current conceptualizations of legal and political notions of rights. Similarly, adversaries utilize Western legal and moral inhibitions against stopping "bad" civil society discourse to undermine the authority and legitimacy of Western institutions and political systems to achieve strategic gains.

Civil society consists of nonstate organizations and informal associations that represent the collective interests of their members. Civil society organizations range from dominant political parties, industry lobbyists, civil rights organizations, and ethnic, racial, and religious organizations to environmental activist groups, registered lobbyists, and so forth. Some groups, so long as they are registered, can advance the interests of foreign governments. Such influence can include Israel's having a lobby group (e.g., the American Israel Public Affairs Committee) representing its specific interests, and then there can be lesser-known lobby groups working on behalf of India, Armenia, Kurdistan, and so forth.[18] Other forms of domestic groups promote local-level interests, such as a neighborhood humane society or industry-specific lobbying groups, such as the pharmaceutical lobby, which spends over $300 million a year.

Characteristics of Western Civil Society

Civil society is vital for holding diverse populations together and is a defining strength of Western liberal democracies. It is liberal in the classical Lockean sense of being a system that highlights and safeguards individual freedoms.[19] Civil society organizations, by definition, are non-state societal organizations that represent citizen interests, playing an important role in checking state power, upholding public interests, and shaping public discourse. Citizens have the right to form nonviolent contractual organizations, helping sustain both economic and political movements.[20] This constructive use of political discourse provides outlets for motivated individuals to pursue their interests in hopes of finding moderate policies and agreements with others, without resorting to destructive behaviors. To paraphrase Sir Isaiah Berlin, in such a moment negative and positive freedoms are balanced; rights are not trampled on either side of the spectrum but remain in constant contestation.[21]

Communal organizations that constitute civil society are seen, from the Magna Carta down, as foundational for liberty and necessary to resist the tyrannical tendencies of unchecked executive power. Communal civil society organizations should be viewed as societal power organizations.[22] They are goal driven and neither benevolent nor malevolent. Some civil society organizations can be utterly uncivil, profoundly illiberal, and easily manipulated if the organizational objectives align with those of a patron(s). Before the Nazi Party took control of Germany in the early 1930s, the country hosted civil society organizations of all sorts, having more Nobel Prize winners than any other country in the world.[23] Unfortunately, many civil society organizations, to include the Nazi movement, happened to be explicitly fascist or contained fascist sympathizers despite Germany's being a highly sophisticated and educated society.

American civil society organizations have played (and continue to play) a crucial role in maintaining pluralist interest representation within the political system, checking the arbitrary impulses of state power at various levels.[24] The strength in America is that, whether one agrees or not, everyone has the same right to organize. In other words, the distinction between a civil society in a liberal democracy and one in an authoritarian system such as China or Russia or an illiberal democracy such as Turkey is that "illiberal" civil society organizations have the same rights as any "liberal" interest group, assuming they remain nonviolent. However, due to laws, traditions, and norms, Western governments cannot investigate the actions of civil society organizations without reasonable cause.

Another fundamental distinction in a liberal democracy is that there are no "gradations" of citizenship. Citizens have the same rights. Naturalized citizens can fight on behalf of their adopted nation, even rising to the highest ranks in public and private life. The late Gen. John M. D. Shalikashvili was a European immigrant from World War II, who moved to the United States at sixteen and learned English watching cowboy movies, becoming the first foreign-born chairman of the Joint Chiefs of Staff.[25] This uniquely American moment was only possible because of the American liberal democratic tradition. In contrast, the average American would struggle to achieve upward social mobility in many other countries, let alone the highest military rank. Nefarious governments, state-affiliated proxies, and nonstate actors can and do exploit these defining characteristics of Western civil society to undermine them from within.

Weaponizing Civil Society

First identified by British anthropologist Gregory Bateson in 1935, "schismogenesis," or the creation of division, refers to a foreign actor's artificial creation of societal rifts or deepening of existing rifts that in turn shape individual behavior. Bateson later utilized his schismogenesis theory in World War II in service of the American government, working for the Office of Strategic Services, an institutional precursor to the Central Intelligence Agency.[26] He relied on counternarrative tactics and deceptive propaganda in the South Pacific to sow disunity among enemy fighters and create schisms in communities supportive of Japanese rule.[27]

The twenty-first-century form of schismogenesis exists within the broader changes in the way people interact and interface across national borders due to globalization, cyberspace, the Internet of Things, and their inherent embeddedness in reality. Before the internet came into supercharged prominence, Robert Putnam identified in his 2000 book *Bowling Alone* a change in social behaviors in America—notably, that Americans were becoming disconnected through a deterioration in collective bonds, cohesiveness, and interactions with one another.[28] Consequently, schismogenesis efforts by adversaries exploit and amplify this trend toward anomie along multiple dimensions and are an embedded element of sociopolitical-information warfare. Issues come from the manipulation of virtual avatars of individuals, domestic lobbying, use of political institutions to hollow them out from within, and the sort of subversion seen during the Cold War. Thus contemporary efforts to attack Western civil society translate into use of various ways and means to deny, degrade, and destroy social capital.

The virtual reality where the virtual avatar inhabits cyberspace has multiple layers; it serves as a human-to-human interface and as a machine-to-machine interface for the Internet of Things. The two are inextricably linked; neither is malevolent or benevolent. Schismogenesis seems easier to generate in liberal democracies due to their inherently permissible legal architecture, which guarantees individual and community freedoms, not to mention the fact that cyberspace provides an easy entry into civil society. There are no rules governing the virtual reality in the West, whereas illiberal governments generate and enforce rules in (virtual) civil society, all under the guise of maintaining societal control. This problem exists because the virtual reality was supposed to be a self-governing reality, the ultimate frontier.

External actors conceptualize accurately this last frontier, the virtual reality that parallels the sentient reality of Western democracies, as an unguarded, undersurveilled, and ill-defined human-to-human interface that can be easily manipulated.[29] This reality is shaped by industry interests with little legal oversight and will remain an unguarded object, as the business models of tech companies necessitate as much. Liberal legal traditions compound the sentiment, and domestic political interests favor the way things are, with powerful lobby groups to reify the status quo. The unguarded virtual reality that everyone is part of serves as an instantaneous, hyperefficient human-to-human interface without boundaries that today constitutes an elemental part of the civic fabric in liberal democracies. The schismogenesis process in virtual reality has four identifiable aspects: tactical depth, illusive intensity, persistence, and scalability.

Tactical depth in the virtual civil society reality is the ability of external actors to directly reach individuals and organizations without compromising their own position and with no fear of countermeasures. Tactics can include direct threats against individuals or organizations, virtual harassment, active denigration of people's reputations, or the feeding of customized disinformation, very much the same way tech companies provide individualized advertisements. The purpose is to generate echo chambers of thoughts and to reinforce certain core ideas and concepts, which can sometimes be injected simply through the use of memes.[30] The introduction of easily customizable and shareable topical pictures with influential phrases (i.e., memes) makes them into effective "thought bombs." Their use contributes to the growing problem of these memes having substantial influence in shaping perceptions of policy-relevant topics, civil society movements, and narratives that run counter to reality. It is a matter of time until machine learning is fused with artificial intelligence (AI) to generate subversive memes that are customized for, tailored to, and deployed against specific individuals and networks of people to maximize strategic effects. It is only natural for such memes to eventually become "precision-guided thought bombs" as sociopolitical-information warfare expands alongside advances in AI and machine learning.

Illusive intensity and *persistence* refer to the reality that people may be subjected to intense disinformation and misinformation, especially over long periods. They may remain unaware of it, but it influences their conceptual frames—much like the old saying "Keep repeating a lie long enough and it becomes the truth." To be able to alter conceptual frames of individuals is to be able to sow discord. It is in this intellectual space that two types of conspiratorial beliefs can be nurtured to the benefit of the foreign actor and the detriment of civil society. The idea or notion can contribute to a monological belief system conspiracy, which generates a serial process of more and more "evidence," providing a breadcrumb trail to the ultimate conspiracy. Then there can be conspiracies within a dialogical belief system, where an internal dialogue within a certain context and an ideological script drive a new conspiracy perception, interpreting facts and evidence through a biased lens.[31] With effective targeting, this can digitally construct a reality for individuals by tailoring their *datascape* (i.e., a visualized environment of data) to their cognitive biases, with the intent of inducing certain behaviors and beliefs that benefit an adversary.

Scalability is simply a by-product of Western citizens' willingness to be surveilled by corporations and have their data sold to foreign third parties. Few, if any, read the legalistic agreements when installing new software on a smartphone. In other cases, incentives are offered, such as discounts and "free" access to applications for surrendering to data collection. The decision to let the interface be self-governed has inadvertently meant that the state (i.e., security agents that are supposed to protect the citizenry) is removed from the process, except under exceptional circumstances. Such surveillance capitalism facilitates predatory data capture and selling that does not worry about sovereignty, let alone externalities related to international security and great power competition.[32] Consequently, much as tech companies generalize the logic of customizing advertisements to fit individual preferences, since they know more about individuals than individuals do themselves, externally malign actors can do the exact same with ease and affordability. These adversarial actions take place with the complicity of the tech companies that are the guardians of the virtual reality that controls and monetizes virtual citizens, precisely because any scrap of data can be viewed through a profit-making lens, just as any bit of data can be weaponized for nefarious purposes. The scalability problem, combined with capitalism, contributes to the weaponization of reality and everything in between.

Individuals, Organizations, and Sharp Power

Henry Kissinger has been a paid consultant to China since he left office. The Freedom of Information Act of 1974 does not apply to private entities; therefore, one is unable to discern the nature of their business relationship. British prime minister David Cameron was immediately added to the Chinese payroll when he left office. Former German chancellor Gerhard Schroeder works for a Russian oil company that is close to Vladimir Putin, and Schroeder is the brains behind the Russian pipeline to Germany, even though everyone highlights the security risks. All these actions are perfectly legal in the West, although former French president Nicolas Sarkozy is still under an open investigation for allegedly receiving campaign financing from Libyan dictator Muammar Qaddafi in 2007. Obviously it is sometimes more convenient if one can hire prominent people to make better business cases. The only problem? It is hard to find a retired Chinese politburo member or Russian KGB agent working as a paid lobbyist for US corporations.

Similarly, an adversarial state recruiting an informer is an act of espionage. However, providing material, ideational, rhetorical, and inspirational support to a community group, lobbying group, or religious organization is squarely within the protected values that define pluralist democracies. Ironically there were many Salafists-jihadists who preached Western destruction in Western cities. They did so because their free speech is legally protected, something not afforded in their tyrannical home regimes. The qualitative difference is significant when adversarial states (and state proxies) utilize the same strengths.

Chinese money continues to have a corrupting influence in Australian politics, and details keep coming out. There is explicit evidence of how Chinese government-affiliated oligarchs have bankrolled the election campaign of a naturalized Australian citizen to become a member of Parliament, essentially making him an agent of the Chinese government.[33] It is important to also pay attention to Communist Party control of business conglomerates in China. This translates into the Chinese parliament consisting of over 100 billionaires.[34] In New Zealand, a naturalized Chinese citizen, formerly a high-ranking military member in a Chinese intelligence agency, is an elected member of Parliament,[35] while his wife, a naturalized New Zealand citizen, runs a civil society organization that explicitly advocates positions favorable to the Chinese Communist Party. In fact, there is substantial evidence to suggest that Beijing has penetrated both political parties in New Zealand, leading allies in the Five Eyes to question New Zealand's membership due to the perception of growing internal influence from Beijing.[36]

Liberal pluralist interest representation and citizenship regimes put these activities in a heretofore-undefined category, and it is difficult to place them within illicit or licit, criminal, and/or counterintelligence domains. These are instances of how external actors have exploited the "civil" in Western civil society, thereby weaponizing it as a battlespace. However, just raising the issue of influence from Chinese money risks accusations of xenophobia. Acting against this influence also raises troubling questions about what counts as subversion and where the boundary between subversion and free expression lies in a liberal society. The Red Scare in the United States in the 1950s serves as a warning that campaigns against subversives can be used to target domestic political opponents instead. Like a sort of autoimmune reaction that runs out of control, the effort to defend the legitimacy and authority of a liberal society against subversion risks diminishing the principles that this effort is supposed to defend.

These complexities of subversion (perceived or real) and flows of information are much less pressing in Russia and China. There are no immigrants, first or second generation, who rise to the highest levels in those countries, but that is not the case in liberal democracies. There are recorded instances of the Chinese state utilizing sharp power tactics to silence Chinese critics and then going on to shape debates using state-sponsored groups registered in liberal democracies, such as New Zealand and Australia.[37] Further to the same point, security agencies in liberal democracies with immigrant traditions neither hold citizens hostages for bargaining purposes nor use the family relations of naturalized citizens to compel them to be complicit in treasonous acts. There is enough evidence to show, and the logic of practice suggests, that Iran, Turkey, Russia, and China are actively trying to leverage the weaknesses that come from transnational familial linkages to make individuals complicit in behavior that they would otherwise not participate in.

New Chinese laws make it obligatory for every Chinese citizen to be an informer on behalf of the Chinese state if asked. Almost all Chinese studying in the West do not want to be spies; they aspire to a job with a tech firm, hoping to make their billions. However, Chinese students (and others from authoritarian states) are exposed to pressure and intimidation from CCP officials, lest they risk the safety of family and friends back home. Such a practice would be the equivalent of the United States constitutionally dictating that every US citizen traveling abroad, at a moment's notice, be willing to work for the state as a spy, with serious repercussions for self, family, and friends for failure to comply.

Chinese law makes it necessary to assume that every Chinese citizen traveling abroad could be a spy. This also includes every company that is run by a Chinese citizen; there is an implicit and explicit expectation written into law by authorities in Beijing that any form of data and information will be turned over to the state when it might benefit China strategically, economically, diplomatically, or militarily. There is no expectation for either foreign or domestic companies working in Western capitals that they can be forcefully compelled to hand over anything without a court-issued warrant, assuming the order makes it past a long, drawn-out legal battle.

Iran and China are increasingly using diaspora populations to influence other countries.[38] Turkey exploits the Turkish diaspora in France and Germany to its own benefit.[39] Russia signals explicit threats with highly public murders, leaving no doubts.[40] This reality is further complicated by the highly charged political atmosphere where the mere suggestion of seeing these issues through a security prism might lead to one being branded as racist. It also appears that fears of espionage from this source, at least in the United States, have no empirical basis. According to a 2021 report of the Cato Institute, individuals of Chinese descent are underrepresented in comparison to the population at large in cases of commercial or official espionage.[41] Despite this reality, these fears create

132 Old and New Battlespaces

suspicions of people of Chinese descent in these societies. Social media warriors funded by Beijing and Moscow will find innovative ways of supporting these fears and any other social movement in the United States and Europe that undermines democratic institutions, hurts the economy, or weakens military power.

The integration of markets and the manipulation or capture of transnational corporations presents another conceptual void in strategy. This problem has security ramifications in a world of big data. By law and tradition, liberal democracies have stringent privacy standards for how much data the government and security agencies can access, but this is not the case in illiberal regimes. Private corporations maintain double standards in their privacy efforts. This is not a value judgment but a reflection of how business decisions respond to incentives.

States structure the domestic markets, generating specific incentive structures. For example, the US government must go through numerous legal hurdles and provisions to access the smartphone of a criminal. However, a smartphone maker provides backdoor keys to the Chinese government, even agreeing to host its data cloud on Chinese government-run servers, in effect collecting and collating data on behalf of the Chinese state.[42] Due to market incentives, these companies become active collaborators with China in supporting the surveillance state, while in the West, the same companies champion themselves as guardians of data privacy. Such shifts point to the future of the virtual reality, where states will assert authority and companies will follow, creating sovereign demarcations. There will be not one internet but many, and states will choose which one they will be part of and the ways in which they can manipulate it in favor of their interests. This is the return of great power politics, but it will be less perceptible because cyberspace is so intangible.

Data Proliferation: The Ammunition to Weaponize Perception

The average citizen in 2020 can generate about 147 gigabytes of data in a day, and Western governments, by law, have almost no access to it.[43] The same standard does not apply to the private sector. They access, collect, collate, and utilize data, selling it as packages to clients. Though US law prevents foreign state-affiliated business proxies of states such as Russia, China, and Iran from accessing such data, enforcement appears to be difficult and inconsistent. In any event, clandestine efforts, such as the Chinese theft of 2015 Office of Personnel Management data, provide a framework against which to match later data.⁴⁴ Data breeches of credit-rating agencies and other companies complement this detailed data about US government employees. China has thereby acquired valuable information about their habits (including those useful for blackmail) and the career patterns of those who become intelligence agents to help undermine US power. The credit-rating agency that gave up credit data on at least 147 million Americans got a slap in the wrist, barely, and American citizens face a hidden burden they did not know about. With commercially available software, the ability to track individuals via their smart devices—where they travel and how they use them—provides a level of high-intelligence accuracy if used as intelligence indicators that Chinese and Russian intelligence agencies would be foolish not to exploit this capability in the new battlespace.

The security implications of credit data (and other forms of personally identifiable data) have rarely been discussed because commerce in the West rarely results in any patriotic fervor toward one's "home" country. This trend becomes increasingly problematic in the rising era of the "Davos Man" and the pursuit of a "home" with the lowest tax burden.⁴⁵ The Davos Man is an extreme representation of the idealized liberal "flat world," where individuals can be here, there, and nowhere, with no loyalty, except to themselves. In the absence of laws, the weakest will always be at the mercy of the most rapacious and ruthless. The Davos Man concept only works if individuals and corporations escape the constraints of sovereignty. Thus far, China and Russia (and other authoritarian states) prove that states are more durable and powerful when exercising authority and power without due regard for the veneer of international institutions that mean nothing more than the socially constructed value lent by the West.

There is the direct infiltration of public debates, interfering with political consensus, and supporting domestic civic society organizations and political parties/candidates. With loose election-finance laws, where individuals are equal to corporations, foreign corporations with proxy firms form super PACs (political action committees) with the express purpose of influencing US policies in their interest. American defense and security agencies are blocked from examining their affiliations, since privacy laws fiercely guard against such efforts. This openness of Western civil society makes it easier to perpetrate schismogenesis. An individual can create a super PAC and advance a political agenda that, in reality, is about advancing the agenda of a foreign agent.

Actors can legally inject and spread ideas of polarity, divisiveness, and fragmentation into free speech debates. Sowing political confusion and anomie in the West gives authoritarian regimes more breathing space domestically and internationally. Besides social media troll bots, Russian backed media and news platforms sow internal division in the West by presenting counternarratives and conspiratorial ideas.[46] Unlike during the Cold War, the United States and its liberal allies no longer employ active measures to defend against such disinformation.

Civil societies can also be put under direct attack, as in Australia and New Zealand. For example, a scholar who identified how China's government was buying political parties and public intellectuals in Australia and New Zealand encountered intimidation and harassment by Chinese agents when she exposed these actions in her published writings.[47] China uses its trade partnerships to punish both countries when their politicians criticize China's policies. These actions show attempts by an adversarial government to "capture" and hold civil society discourse hostage in pursuit of specific political ends in other countries.

There is evidence that Russia supports various civil society groups in the United States and elsewhere in the West. For example, Russia appears to have funded environmental groups (e.g., antifracking groups) and the National Rifle Association (NRA).[48] Such actions by Russia to "protect" the environment and "support" the NRA are not virtuous. Instead, such support to environmental groups protects Russian economic interests and support of the NRA allows Russia (and similar authoritarian governments) to paint American democracy as a dangerous experiment that should not be emulated.[49]

Actions by foreign entities to support other civil society groups indicate that American politics are being subverted with an intent to foment long-term instability, with the duality of propaganda focused on promoting fascist and Communist views on both sides of the political spectrum. If one accepts the idea that civil society groups are designed to uphold the rights of citizens, then one should also assume that America's adversaries believe that idea too. China and Russia find it in their national interests to fund and support "hot-button" civil society groups for the purposes of creating downward spirals of societal tension and violence, which fits the model of schismogenesis.[50] This problem has been best exemplified by Russian troll farms creating "American as apple pie" movements centered on amplifying a data point and/or anecdotal evidence and then attempting to make it into its own story or social movement. This is similar in principle to astroturfing, whereby certain messages are sponsored by undisclosed individuals but appear to be a grassroots movement. The process then gains momentum and legitimacy as bounded rationality compels caveman instincts and tribalism, resulting in most people choosing one side or the other.

With the advent of Bitcoin (and similar cryptocurrencies), covertly funding various civil society groups has become easier, as well as more difficult for security agencies to detect and defeat.[51] Moreover, individual citizens who already espouse conspiratorial views can be clandestinely funded by foreign entities, helping these conspiracy theorists elevate and amplify their platforms, expanding their power and reach. In such a milieu, a range of provocateurs, be they genuine home-grown individuals or foreign posers, bring a new sense of normalcy to internet behavior, despite its being highly extremist and divisive in nature.[52] Influence operations (i.e., "psyops") are so constant that they become normalized behaviors, altering the tenor of conversations and debates, with the intensity of an argument being more important than its actual merit or effectiveness as a policy.[53] Political parties in the West inadvertently encourage these actions and behaviors by fostering a toxic information environment, full of culture wars and propaganda, because it suits their short-term political interests without due regard to strategy and considerations of long-term ramifications.[54]

The development of AI will make it easier for social media warriors to establish more false social media start-ups and supporters in the era of constant influence operations. AI bots can operate thousands of social media accounts to interact in a humanlike fashion with citizens, recruiting real humans to support their false causes. Machine learning may amplify this weaponizing of civil society processes, as it will enable the exponential targeting and dispersing of negative civil society interactions via social media networks by customizing and tailoring messages in a way that appears genuine. This form of computational propaganda will accelerate as algorithms work in tandem with machine learning—with some human curation—to generate dynamic programs that can interact online with other users in virtual civil society.[55]

Social science advances will augment this as well, developing an understanding of what triggers certain behaviors and motivations and then micro-tailoring messaging and information to maximize effects. One might worry with good reason about living in a simulation like *The Matrix*, as such technological advances paired with social science in the not-so-distant future will create false realities, tailored to and customized for each citizen. This weaponization of perception can be pursued at a much lower cost than building expensive weapon systems.

Schismogenesis against Western civil society contributes to the undermining of democratic institutions, thereby complicating the policymaking process. It is a cost-effective asymmetric strategy that weakens Western power and strength, without requiring substantial investments in conventional armaments. Maintaining conventional deterrence and combat primacy may necessitate investments in advanced weapons systems, but such investments must be aligned with investments in counterstrategies against cheap social media warrior attacks on an unguarded civil society.

In sum, authoritarian adversaries have effectively turned democracy, globalization, and capitalism against the liberal democracies. They have done so with the complicity of citizens and private enterprises in these societies, whose political and legal institutions become tools for external adversaries. This calls into question seeing warfare in terms of the Clausewitzian trinity—government, people, and army—meaning that civil society, a sinew of the three, is overlooked as a defendable position.[56] Clausewitz, at the time of his writings in the early nineteenth century, was not attuned to the concept of a robust civic-minded society as the glue binding the citizenry, military, and government to one another. Attacking civil society appears to be more successful than trying to target each part of the trinity.

Conclusion

The world stands in stark contrast to what most imagined at the end of the Cold War it would look like. Industrial capitalist democracies that were meant to chart the course of humanity staring at the horizon at "the end of history" face serious threats from within and without.[57] Somewhere along the way, in the transition to an information-age economy, Western political leaders facilitated institutional rot for narrow self-serving purposes. Worse, they abandoned principles and strategic views of the world that made the United States and Europe so preeminent in 1991. Dictatorial and illiberal states have found ways to leverage geopolitical realities, Western strategic contradictions, and emergent domains to their advantage.

Western militaries now recognize—with the great Global War on Terror distraction of two decades of counterinsurgency and counterterrorism behind them—that adversaries have learned to incorporate technological advances such that they have managed to compress, converge, and expand traditional battlespaces. Simultaneously, they have developed their abilities to leverage the emergent domains of space, cyber, and civil society, turning them into a new battlespace defined by *schismogenesis*, where a sociopolitical-information battle rages in capturing the high ground of individual perception—a battle that the United States and its partners are losing.

Notes

1. VENONA (archive), National Security Agency, www.nsa.gov/news-features/declassified-documents/venona.
2. Canada has several broadcasting provisions and laws that can result in a news outlet having its license revoked for reporting fake news. Since 2018, the EU has had an action plan for dealing with fake news: "Tackling Online Disinformation," European Commission, https://ec.europa.eu/digital-single-market/en/tackling-online-disinformation.
3. Czosseck and Geers, *The Virtual Battlefield*.
4. The absurdity of not reading the fine print when accepting the Itunes terms of agreement was best satirized by *South Park* parodying the exploitation that many tech companies engage in. Trey Parker, dir., "Humancentipad," *South Park*, Season 15, Episode 1, aired April 27, 2017.
5. "Mobile Fact Sheet," Pew Research Center: Internet and Technology, June 12, 2019; Denis Metev, "39+ Smartphone Statistics You Should Know in 2020," *Review 42*, November 21, 2020.
6. Friedman, *The World Is Flat*.
7. "Clinton's Words on China: Trade Is the Smart Thing," *New York Times*, March 9, 2000, www.nytimes.com/2000/03/09/world/clinton-s-words-on-china-trade-is-the-smart-thing.html.
8. Permission was granted to reprint part of this article: Jayamaha and Matisek, "Social Media Warriors."
9. Fink, "Dangerous Speech, Anti-Muslim Violence, and Facebook in Myanmar"; Ibrahim, *The Rohingyas*.
10. "Social space" is a term used to describe the ability of different people and groups to collectively organize, sometimes against the government. Reno, *Warfare in Independent Africa*, 9.
11. Malcomson, *Splinternet*.
12. Geremie R. Barme, Sang Yegeremie R. Barme, and Sang Ye, "The Great Firewall of China," *Wired*, June 1, 1997.
13. Scoggins, "Propaganda and the Police."
14. Chen, "Political Context and Citizen Information."
15. Qiang, "The Road to Digital Unfreedom."
16. Kuehn and Salter, "Assessing Digital Threats to Democracy, and Workable Solutions."
17. Mazarr, Bauer, et al., *The Emerging Risk of Virtual Societal Warfare*.
18. Mearsheimer and Walt, *The Israel Lobby and US Foreign Policy*; Ashok Sharma, "Behind Modi: The Growing Influence of the India Lobby," *The Conversation*, June 27, 2017; Ömer Taşpınar, "The Armenian Lobby and Azerbaijan: Strange Bedfellows in Washington," *Brookings*, March 8, 2010; Eric Lipton, "Iraqi Kurds Build Washington Lobbying Machine to Fund War Against ISIS," *New York Times*, May 6, 2016.
19. Locke, *Locke*.
20. North, Wallis, and Weingast, "Violence and the Rise of Open-Access Orders."
21. Berlin, *Four Essays on Liberty*, 162–166.
22. Mann, "The Autonomous Power of the State."
23. Herbert, "Berlin," 75–83; Berman, "Civil Society and the Collapse of the Weimar Republic"; Mann, *Fascists*, 162–205; "Nobel Laureates and Country of Birth," Official Web Site of the Nobel Prize, June 11, 2018.
24. Skocpol, "Civil Society in the United States," 109–121.

25. Shaila Dewan, "Gen. John M. Shalikashvili, Military Chief in 1990s, Dies at 75," *New York Times*, July 23, 2011.
26. Bateson, "Culture Contact and Schismogenesis"; Lipset, *Gregory Bateson*, 143–144.
27. Price, "Gregory Bateson and the OSS"; Price, *Anthropological Intelligence*, 239–242.
28. Putnam, *Bowling Alone*.
29. The concept of social media warriors leveraging cyberspace alongside the emergence of a new battlespace was first conceptualized in 2019: Jayamaha and Matisek, "Social Media Warriors."
30. Zakem, McBride, and Hammerberg, *Exploring the Utility of Memes for U.S. Government Influence Campaigns*.
31. Goertzel, *Turncoats and True Believers*; Goertzel, "Belief in Conspiracy Theories."
32. Zuboff, *The Age of Surveillance Capitalism*.
33. Philip Wen, "China's Patriots Among Us: Beijing Pulls New Lever of Influence in Australia," *Sydney Morning Herald*, April 14, 2016.
34. Sui-Lee Wee, "China's Parliament Is a Growing Billionaires' Club," *New York Times*, March 1, 2018.
35. Jamil Anderlini, "China-Born New Zealand MP Probed by Spy Agency," *Financial Times*, September 13, 2017.
36. David Fisher, "Chinese Communist Party Link Claimed," *Otago Daily Times*, May 26, 2018; Eleanor Ainge Roy, "New Zealand's Five Eyes Membership Called into Question over 'China Links,'" *The Guardian*, May 27, 2018.
37. Elif Selin Calik, "A Newly Coined Phrase: 'Sharp Power' and Reasons for Attributing It to China," Rising Powers in Global Governance, January 6, 2018, https://risingpowersproject.com/a-newly-coined-phrase-sharp-power-and-reasons-for-attributing-it-to-china.
38. Timothy Heath, "Beijing's Influence Operations Target Chinese Diaspora," *War on the Rocks*, March 1, 2018; Robinson, *Modern Political Warfare*.
39. Mencutek and Baser, "Mobilizing Diasporas."
40. Lucy Pasha-Robinson, "The Long History of Russian Deaths in the UK Under Mysterious Circumstances," *Independent*, March 6, 2018.
41. Nowrasteh, "Espionage, Espionage-Related Crimes, and Immigration."
42. Sherisse Pham, "Use iCloud in China? Prepare to Share Your Data with a State-Run Firm," *CNN*, January 11, 2018.
43. In sum, approximately 2.5 quintillion bytes of data are being created every day by humans. "Data Never Sleeps 6.0," Domo: The Business Cloud, 2018.
44. Internally, US intelligence officials became aware of the Office of Personnel Management (OPM) hack in 2012, but it was not openly disclosed until 2015. That is when some evidence arose to suggest that China may have secretly altered some OPM files to leverage specific individuals with national security jobs. Zach Dorfman, "China Used Stolen Data to Expose CIA Operatives in Africa and Europe," *Foreign Policy*, December 21, 2020.
45. Huntington, "Dead Souls."
46. Allen and Moore, "Victory Without Casualties."
47. Matt Nippert, "University of Canterbury Professor Anne-Marie Brady Concerned Break-Ins Linked to Work on China," *NZ Herald*, February 16, 2018.
48. Merill Matthews, "Democrats Dig for Russian Connection and Uncover Environmentalists," *The Hill*, October 26, 2017; Tim Dickinson, "Inside the

Decade-Long Russian Campaign to Infiltrate the NRA and Help Elect Trump," *Rolling Stone*, April 2, 2018.

49. Isaac Stone, "How Chinese Media Covers U.S. Gun Violence," *USA Today*, February 17, 2018; Erin Griffith, "Pro-Gun Russian Bots Flood Twitter After Parkland Shooting," *Wired*, February 15, 2018.

50. Philip Ewing, "Russians Targeted U.S. Racial Divisions Long Before 2016 and Black Lives Matter," *NPR*, October 30, 2017.

51. Jahara Matisek, "Is China Weaponizing Blockchain Technology for Gray Zone Warfare?" *Global Security Review*, March 13, 2018.

52. Stanley, *How Fascism Works*.
53. Pomerantsev, *This Is Not Propaganda*.
54. Stanley, *How Propaganda Works*.
55. Woolley and Howard, *Computational Propaganda*.
56. Clausewitz, *On War*, 32, 89.
57. Fukuyama, "The End of History?"

7
New Battlespaces and Strategic Realities

LIBERAL DEMOCRACIES TODAY CAN NEITHER CLAIM STRATEGIC PRImacy nor project power unchallenged. Though the United States and the North Atlantic Treaty Organization (NATO) managed to win every tactical engagement in the last two decades of counterinsurgency and counterterrorism, none of those tactical gains have been turned into strategic advantages, much less strategic primacy. Many more citizens in liberal democracies, including in the United States, question basic precepts of their own political systems. Old party systems have collapsed in some of these countries as new identity-based and nativist parties have come on the scene. These new politicians practice what political scientist Adam Przeworski calls "subversion by stealth." This involves the use of existing institutions outside established norms to weaken basic elements of democracy, such as competitive elections, rights of free speech and association, and the rule of law. This happens in steps through a new intensity of instrumental manipulation of voting rules, vilification of media, and so forth. There is nothing "undemocratic" or unconstitutional about this process, but it generates widespread feelings of mistrust and reprobation toward politicians on the part of most citizens.[1]

These developments reflect a shattering of the ideational consensus that animated the US and NATO strategic focus that survived to the end of the Cold War. They also create an internally generated sort of schismogenesis, a situation that provides adversaries with a wide range of opportunities for what the counterintelligence community best describes as "recruitment" in these divided political systems to achieve the adversary's intelligence objectives. Classic examples from long before the current situation include help finding "dirt" on rival politicians and passive coordination in pushing preferred narratives, actions that do not necessarily rise to the level of crimes. As the last chapter emphasized, narrowly focused influence operations can coordinate assistance to "recruits" who have no contact with their beneficiaries and may not even know of the origin of the assistance. Aided now by changes in the GRINS framework of technology, organization, and the character of political regimes, these engagements would have been beyond the pale in the old Cold War strategic environment.

Transformed Battlespaces

In the current complex strategic context, the armed forces of the United States and its NATO allies recognize that the battlespaces are now irrevocably *expanded, converged,* and *compressed.* Simultaneously, traditional war-fighting domains and emergent domains keep integrating, creating great strengths and vulnerabilities. The confluence has given rise to the reality of synchronous multidomain operations.[2] For example, a simple kinetic enterprise, such as a time-sensitive target-acquisition mission, involves blowing up a door and capturing a particular actor. In the multidomain era, this "simple" action now involves cyber, space, air, and land, and depending on the time and place and the infiltration and exfiltration method, it might require maritime assistance.

This integration of domains and the resulting jointness have helped overcome tactical blind spots and supported remarkable tactical proficiency. However, since these advancements take place within (and are shaped by) domestic legal frameworks, there are many doctrinal gaps that remain unaddressed, since those gaps are decidedly kept outside ongoing strategic discussions due to ethical, moral, legal, and political concerns inherent in liberal democracies. These concerns reflect core values of open democratic systems. They also continue to be inflected by the unusual circumstances of the unipolar moment that allowed US officials and the country's foreign policy establishment to deal with the rest of the world on their own terms. The absence of adversaries that could test American will permitted the optimistic extension of these legal and normative values in ways that now hinder US competition in this transformed battlespace. Adversaries' capacities to influence the course of internally generated divisive political debates contribute to the difficulties of responding to new challenges and ultimately to articulating a clear and widely accepted strategic purpose.

Assuming away strategic blind spots out of moral, legal, and political expedience has strategic implications when adversaries turn them into critical vulnerabilities. This reality is captured in the way competitors view the current operational environment in contrast to the United States and its allies. That is one of the fundamental functions that explains strategic failures of the United States, NATO, and other allies with open liberal societies. Hostile actors cannot compete with their conventional militaries at the moment, but they readily identify points of weakness and seize opportunities across this wider spectrum of competition to challenge American power. This is a problem of political will and the West's approach to the operational environment. Improving strategic competence requires recognition of the true contours of this change in the character of warfare, an outcome that this book aims to promote. It is important to recognize that moving beyond this strategic narcissism also entails a complicated political discussion about the aims of American power in the world, the trade-offs between core values that should not be compromised (some of which adversaries identify as tools to be used against the United States), and the actions that need to be taken to respond to this new strategic environment.

Peer and near-peer adversaries of the United States view the strategic milieu as complex but also as *integrated* and *synchronous*. Actual and potential adversaries accurately picture the altered battlespaces as integrated, and when they settle on an end state, they tend not to perceive moral, legal, or political inhibitions if they are interested in marshalling the full range of tools to operate in interstitial spaces and pursue asymmetric operations against US power. From that adversary's perspective, sociopolitical domains are part of the strategic environment that cyberspace and other social and technological developments have brought closer to their grasp. In essence, that which US and allied officials would have recognized as political warfare during the Cold War has returned—though with a much more diverse and complex character.

Consequently, upon deciding on a political and military end state, adversaries operate synchronously across domains and gain tactical advantages cross-synchronously. These contrasting views of the operational environment—the new battlespace—are represented schematically in Figures 7.1 and 7.2.

Figure 7.1 American View of the Operational Environment

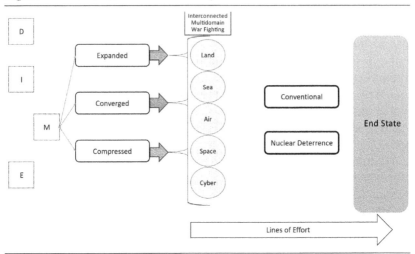

Figure 7.2 Adversarial View of the Operational Environment

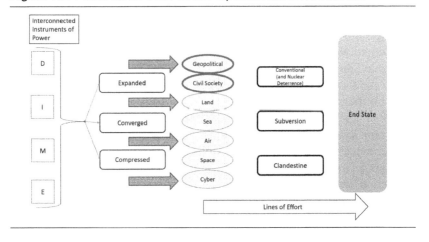

Integrated and Synchronous Strategic Environment

In the conventional US strategic formulation, no matter what the desired strategic end state, the pursuit of that objective must be done without compromising democratic ideals. To do otherwise would be self-defeating. If citizens in whose name these militaries fight do not support the political ends of the fight, if political leaders cannot articulate clear ends, or if citizens have qualms about the ways and means that are used to execute that fight, it will become impossible for political leaders to leverage all of the instruments of national power to pursue US interests. Regime type then shapes strategy, just as it shapes warfare.

If the acronym DIME (diplomatic, informational/ideational, military, and economic) broadly captures the instruments of national power, their application is separated by design in the United States and other democratic countries. The *M* in DIME is broadly defined to include the entirety of the national security architecture: the military and an array of intelligence organizations, which are externally focused and retain no domestic remit. Therefore, the strategic conception of the current operational environment by the armed forces takes place within narrowly defined scope conditions as shown in Figure 7.1. By design, there is a kinetic premise (someone must be shooting something) in this strategic formulation in order to identify aggression. Simultaneously, prescribed scope conditions direct military thinking outward, while restricting that thinking to the land, sea, air, space, and cyber domains. This contrasts with the conception of adversaries.

The *D*, *I*, and *E* in DIME are separate autonomous lines of effort, just as *M* is its own effort. In an ideal world, the four are supposed move in unison, while the military limits its concern to the *M* in DIME. This is a logical extension of democratic principles. A contrasting example focusing on information operations will better illustrate the challenge.

The US Armed Forces, including the Cyber Command, Special Operations Forces community, and intelligence organizations, retain the capabilities to conduct information operations to change perceptions at all levels. However, they only conduct information operations as part of a broader campaign, where the influence operations remain within predefined scope conditions and timelines, and they never conduct shaping operations against their own citizens. The introduction of the US Space Command and Space Force suggests the need for further integration in maintaining the "high ground" across all domains in the worsening era of sociopolitical-information warfare.

In contrast, adversaries prioritize information and shaping operations as a consistent strategic line of effort; in doing so, they build on their economies-of-scale approach. These authoritarian governments already spend significant time, effort, and resources conducting information operations against their own citizens. This capacity allows them to expand their operations internationally with little cost. Put differently and explicitly, armed forces in liberal democracies are not used as a tool to control and suppress the people but are designed to protect them. On the contrary, the armed forces and the entire security architecture of China, Russia, and Iran, for example, are designed to control and suppress their own people to protect the regime. Controlling their respective virtual and sentient civil societies and engaging in constant information operations to control the perceptions of their subject-citizens remain core military functions. Similarly, they do not consider themselves bound by any international norms or conventions since undermining these is a core objective. That generates a contrasting view of the operational environment.

Cross-Domain Synchronicity in New Battlespaces

Strategic adversaries are intimately familiar with a battlespace that is expanded, converged, and compressed. However, the nature of authoritarian regimes makes them better situated to make sure all instruments of national power (e.g., DIME) are better aligned and unified to execute a whole-of-state strategy toward a desired end state. Their multidomain operational environment is not restricted as it is in the Western formulation. In the adversarial view (see Figure 7.2), the land, sea, air, space, and cyber domains remain critical. However, when it comes to contesting status quo powers, competitors correctly recognize that to engage in interstitial tactics, they must *broaden their scope of targets*, an enduring feature of the asymmetric nature of warfare in similar geopolitical struggles through history. But now unguarded virtual civil societies become targets. The virtual avatars of every individual reside in this cyber ether and are at the mercy of, and have been monetized by, private enterprises, where myriad geopolitical realities (e.g., international organizations, institutions, laws, and norms, etc.) are a part of the operational environment. This view of the world is fundamentally at odds with most Western conceptions of what war and competition are believed to be today.

Integrated Multidomain Operational Environment

For the adversary focused on symmetric modes of challenge and operating in the integrated, synchronous multidomain operational environment, nothing is sacrosanct, and everything is open for contestation. This is logically consistent from the point of view of revisionist powers, as it has been through the various changes in the character of warfare surveyed in this book. That is, strategy is about creating desired effects, and adversarial civil societies and the geopolitical realities at particular moments are all ultimately power realities that impact and shape state behavior. It makes sense that adversaries would see this environment as designed by and for the benefit of the United States and its close allies, representing the realities of power at the time they were devised. International norms, laws, and transnational justice leagues are seen as extensions of the ideals of dominant powers with universalistic pretension that in the end seeks to undermine illiberal regimes and dictatorships. Indeed, American political leaders in recent decades to the present frequently have stated that this is a paramount US strategic interest. While the United States and others can believe they are under duress from hybrid warfare and novel tactics, illiberal and authoritarian adversaries view the behavior of the United States and its allies in the post–Cold War era as the original version of hybrid warfare that sought to undermine their rule, authority, and sovereignty. This is an accurate assessment of actions taken by the United States and its various coalitions since 1991.

China, Russia, and other actors with revisionist aspirations did not view the deepening and expansion of the US-based global order after the Cold War as advantageous. The deepening of the US-centric liberal order was interpreted as a nonkinetic form of creating strategic effects against the nondemocratic states. While most people in the United States viewed the growth of an American-led order as peaceful and beneficial for all, the targets of this US-led order perceived this as a new form of disruptive warfare meant to undermine their autonomy, geopolitical realities, and social control. The many international nongovernmental organizations (NGOs) and their employees were interpreted as subversive instruments and agents. The Russian foreign-agent law, enacted in 2012, required organizations that receive foreign donations to register as "foreign agents" if they were deemed to engage in political activities inside Russia. The list of targeted organizations includes Human Rights Watch, Amnesty International, and Transparency International, each of which has been accused in state-influenced media of advancing the political interests of the United States. Russia's undesirable organizations law bans NGOs deemed a "threat to the constitutional order and defense capability, or the security of the Russian state." Undesirable organizations include the Washington-based National Endowment for Democracy and International Republican Institute, both of which had been active in democracy-promotion activities in Russia that Russian authorities essentially identify as subversion.[3]

Individuals in these geopolitical realities, such as prosecutors attached to international tribunals, operate with a great degree of autonomy and relatively little concern for shifts in global power relations. That is to say, if one believes that universal jurisdiction is a concept that is autonomous from the unipolar moment that fostered its extension to defendants from the former Yugoslavia and various conflicts in sub-Saharan Africa, the end of that unipolar moment should not matter greatly. But these judicial actors and other arbiters of international affairs like them are political actors embedded in the broader geopolitical context—and the context has changed. But even in the unipolar moment, international tribunals were able to operate effectively only when applied to the world's weakest and most strategically marginal actors, such as former warlords, deposed presidents of sub-Saharan African countries, and the losers of wars in the Balkans that ended with NATO military intervention. In contrast, the Donald Trump administration responded to International Criminal Court (ICC) suggestions that it would investigate US soldiers for actions in Afghanistan with threats to revoke visas and even suggested it would arrest ICC officials in some circumstances. The Joe Biden administration watered down its language but abides by the same reality.

From a rhetorical point of view, members of the transnational justice leagues see accountability as reserved for the accused, for as prosecutors they are seen to be above politics and on a universalizing mission. These individuals are very much cut from the same ideational (if not ideological) cloth as the US officials who authorized the US invasion of Iraq in 2003 because they thought this military action would bring Iraq's people the benefits of universal values of justice against tyrants and democratic peace. From the point of view of adversaries, both are grand gestures of naivete. If some people in the United States and Europe are idealistic enough to assume that these transnational power realities converge on some sort of universal end state and their employees are noble, apolitical actors, therefore both are sacrosanct and above politics, that is all for the better. From the point of view of adversaries that seek to alter the status quo, this reality amounts to multiple loci of power that are defenseless and open for co-optation or coercion.

Now the costs of a lack of strategic vision begin to mount. The Covid-19 pandemic is a case in point. The global health crisis caused by Covid-19 originated in Wuhan, China. However, the UN Security Council did not hold a meeting about it, because China chaired it in 2020. Ignoring the Covid-19 crisis was not merely intended to shield China from scrutiny; the apathy contributes to the undermining of the United Nations, which is an important objective for an increasingly economically powerful China. The United Nations, created at the end of World War II, is the most prominent symbol of Western power. It is better for revisionist and revitalized powers to neuter it (and similar post-WWII international institutions in which the United States predominates) so that in some cases they can create new ones. Another option is to operate within existing institutions to redirect the focus of their activities so the common goods they produce are associated with the adversary. This is a costly strategic irony. The states that built and bore the greatest cost to uphold these realities have handed them off, putting them at the mercy of spoilers that ultimately seek to undermine the very order that these institutions created.

Adversarial weaponization of civil society and international institutions parallels the naive understanding of where geopolitical realities come from and why international institutions and norms are able to function. By design, liberal democracies have made the virtual reality where a virtual avatar resides in an unguarded space and where adversaries can (and do) reach individuals directly. It is a borderless virtual reality, though, as noted in the previous chapter, politicians and regulators in the United States and in other countries have identified this exposure as a problem. But citizens of countries with open societies remain exposed to influence by adversaries. The weaponization of reality and everything in between is a by-product of this process because technology and data enable precise targeting and intensity to amplify whatever objectives might be pursued by an adversarial state. This points to how adversarial states increasingly pursue strategic interests without provoking a military response, essentially neutralizing notions of violence as strategic communication and signaling.[4]

As a consequence, it is hard to imagine how one constructs a relationship of deterrence between the United States and its adversaries within this expanded realm when the adversary has the potential to exercise direct influence over citizens (including, as noted previously, politicians and business leaders) in a targeted state. This situation highlights the conundrum for the United States and other countries that benefit from a globalized system of commerce and ideas, yet find that they have to construct barriers that limit the scope of that system if they are to protect that system and their places in it.

The critical element of this novel operational context is that the unipolar moment's concept of universality means that US political leaders have tended not to achieve victories outright (i.e., the transformation of Iraq and Afghanistan) because failure in these efforts ultimately has not been seen as harming national interests, at least not in the fundamental way that defeat in World War I or II would have been interpreted. Does it matter if the United States was defeated in Afghanistan? Probably not, if one's reference points are US public opinion and demands on US financial and material resources. Indeed, admitting defeat earlier is prudent.

But adversaries still exist. Admitting the defeat in Afghanistan that even very cursorily informed people (and, one suspects, many a high official) recognized years ago is a good idea. But doing so without attention to the strategic environment is a bad idea. Adversaries adopt an interstitial strategy of slow, smooth, and steady accumulations of seemingly little, even negligible, discrete accomplishments over time and across space, which in their cumulative manifestation generate strategic realities. A core proposition of the argument in this book is that these strategic realities matter—that the unipolar moment did not mean that the enduring fundamental nature of warfare and strategy went away. Losing a little in one place invites these probes. Directed-energy attacks on US troops is one instance of an interstitial activity that reaps no tactical or operational advantage but moves that strategic reality ever so slightly, an effect that is magnified when there is no coherent response.[5]

These adversaries win by not losing, and they will not lose because they will not fight a conventional war outright. Status quo powers do not have to be defeated in bloody battles to lose their strategic advantage. They lose when the status quo changes, leading to the emergence of novel strategic realities. At the hour of their triumph, the United States and a large part of Europe—those areas not under the domination of the Soviet Union up to 1989—forgot the effort it took to stay the course that led to their victorious emergence at the end of the Cold War. Instead, many of the beneficiaries of this development assumed the automaticity of geopolitical transformation, when in reality any transformation is about power with distributional consequences; it is therefore never automatic but only contingent on decisions and the convictions of the few who will assume risks and act on their convictions. Consequently, capitalism, globalization, technological advances, and all the well-intentioned transnational realities are being utilized (i.e., weaponized) by nondemocratic adversaries to undermine liberal democracies from within and without, inside and out.

Political expressions cast as populist movements in liberal democracies reflect the shifting ideational foundations. Adversaries increasingly view such movements as mere opportunities to widen the scope of their maneuver space in the multidomain operational environment, where discrediting the system itself, and doing so as an end in and of itself, is a strategic line of effort. The level of suspicion aimed at Western institutions by their own people is a strategic irony in this context precisely because robust democratic institutions were a strength during the Cold War, and leaders in democracies treated them as being tied to the survival of the West against Soviet strategic designs.

Challenges of Automation, Artificial Intelligence, and Big Data

The United States and other states generally aligned with it retain a competitive edge in technology. The challenge posited here does not refer to maintaining the technological edge, for economic incentive structures shape the innovative spirit in capitalist democracies, and that is beyond the scope of this discussion. The challenge refers to managing the duality inherent in scientific knowledge in its applied manifestation. Scientific knowledge turned into technological applications is as infinitely benign as it is infinitely malign. Armed forces and security services in the United States and other democracies face a daunting challenge in harnessing advanced technologies, on the one hand, and defending the innocent who rely on advanced technological adaptations from malicious actors, on the other. This challenge in the future will be even greater than the problems of adversaries leveraging social media and turning civil society into a new battlespace.

Many concerned officials have come up with ad hoc measures as challenges rise. However, the pace of technological change is so rapid and the inherent unaddressed vulnerabilities inherent in technological change are such that there will come a day when strategic logic will compel liberal democracies to develop collective measures to harness technological advances, build collective defenses, and devise collective offensive strategies to leverage them against adversaries. Prospects for advances in automation, artificial intelligence (AI), data generation, machine learning, and quantum computing, to name a few, present the daunting nature of the emerging challenge. However, the challenge of quantum computing remains different from the rest. This is because such technology will enable quantum sensing in all domains, meaning that a submarine will no longer be able to hide its movement and a stealth F-35 will be detectable. Moreover, whichever country establishes quantum computing will establish primacy in unhackable communications while being able to easily decrypt and decipher those of other states utilizing nonquantum communication equipment.[6]

Technological advances take place at an ever-faster rate (using patent apps worldwide as a proxy), and China holds the fastest rate of change.[7] By the end of 2019, China surpassed the United States in international patent applications with 58,990 Chinese filings versus 57,840 American filings.[8] Technological change in a country is shaped by its underlying political economy and the resulting incentive structures in place. Market signals drive tech changes. Two defining "market makers" in the United States, historically, have been the Department of Defense, followed by the space agencies—the US Air Force remains the largest space entity. Historically, private entities solved defense puzzles in terms of products, software, weapons systems, aircraft, carriers, tanks, and the like. Advances in defense technology usually also generate commercially viable positive externalities. The logic is now reversed.

A qualitative difference in postindustrial technological advances reverses the usual logic whereby armed forces often attempt (with the exception of very specialized areas, such as stealth technology, which has little commercial viability) to incorporate advanced technology that is already publicly available. The private sector holds the keys to most advanced technology. But in liberal democracies, commercial logic, national interests, and defense-offense concerns do not always align, and for reasons of public opinion or employee concerns, private companies may refuse to work with defense entities, as happened in the case of several Silicon Valley companies—even though the same companies were providing services to illiberal and authoritarian states. Yet the vulnerabilities created by advances in technology get compounded by their integrated nature, making public-private collaboration necessary, if not imperative.

The critical factor will be the capacity to adapt organizationally to use the technology in military affairs, as when Germans developed blitzkrieg warfare. The United Kingdom and France developed tank technology, and thinkers like Alfred Thayer Mahan and Liddell Hart conceptualized the organizational starting point, but these governments did not take advantage of their head start. The United Kingdom developed the first aircraft carriers but used them to support spotter aircraft as adjuncts to battleships. They were too attached to the big weapons systems to which they had become accustomed, a problem familiar to anyone who has followed the saga of the F-35, a $1.6 trillion program begun in 1992 based on a strategic rationale suited to fighting the Soviets, which was supposed to be an affordable fighter to serve air force, marine, and navy needs. Now AI technological and organizational competition will be multilateral and more diffuse than Cold War competition. Unlike Cold War missile technology (and, to some extent, early tank technology), AI development is a commercially driven enterprise in the United States, and its diffusion incentivizes mimicry among competitors keen to catch up. Collaboration will be challenging within this organizational and cultural environment—very different from what US Department of Defense personnel are used to. Academics and start-up tech entrepreneurs have little patience for the extremely slow, bureaucratic pace of that federal agency, a problem that adversaries may be able to address with more finesse.

Automation Is Not the Future; It Is the Present

Automation, at least in its advanced form, and artificial intelligence mutually reinforce each other, but they are contingent on data. The more data there is, the better both will be. The advent of 5G (and 6G+ and beyond) networks and interconnected, embedded technologies, together with the linkage of even the most mundane activities to smart devices, translates into every individual, entity, and device generating ever-increasing amounts of data with more fidelity. As more individuals are turned into data-generating products that companies monetize, people's data will be more widely proliferated, making it all too easy to weaponize.

Continual advances in AI, automation, machine learning, and much more all center on the sharing of data for optimizing product efficiencies. However, none of the private entities are legally bound to publicize instances when their networks are either hacked or individual data is compromised by either state or nonstate actors. The fear is twofold. It is well known that companies hacked by Chinese entities often fear publicizing as much, knowing they might face economic retaliation from China. Desperate to maintain market access, some of these companies become complicit with the illicit activities of the Chinese state.[9] Hiding the fact that people's data is compromised also makes business sense, since no company wants to attract negative publicity voluntarily. As a result, much of individual data that gets compromised—be it from banks or credit agencies or just related to little children using tech in schools—never gets reported precisely because there is no law that requires it to be. Therefore, private entities can monetize individuals and are rarely held accountable for losing their data. This is a vulnerability that external adversaries exploit on a daily basis, and the danger will increase as advances in technology integrate people, just as adversaries leverage novel battlespaces to generate adverse effects in liberal democracies to undermine them from within, raising a dual challenge.

Major technological advances, in a laissez-faire economic environment, transform people's lives for the better. Therefore, it should be a strategic priority to reify the incentives and institutional configuration to maintain or advance the pace of innovations. However, it will soon become impossible to disregard the inherent vulnerabilities that these technological advances generate. Today's minor sector-specific challenges will eventually morph into broader strategic challenges. This is because—much as Stanley Baldwin told the British Parliament in a 1932 speech, "The bomber will always get through"—the twenty-first-century hacker will eventually get through too. While countermeasures eventually arise to stop the bomber—or cyber warrior—it becomes difficult to stop the constant, ever-persistent reality of sociopolitical-information warfare, especially as foreign actors pursue schismogenesis and prey on cognitive biases, domestic politics, and personal grievances.

Adversaries that are already operating within the strategic paradigm of a cross-synchronous multidomain operational environment will find emerging vulnerabilities to be a great gift. In that moment, the solution will not lie in heavy-handed regulatory oversight and legislative control. Rather, the solution will lie in a legislatively mandated collaborative approach between the private and defense sectors with judicial oversight to prevent overreach by either party.

Very soon, the United States and many other open societies will need to have the tough conversation about the juridical boundaries and traditional divisions of labor that have shaped the protection of citizens in liberal democracies. Dangerous speech, akin to yelling "Fire!" in a crowded movie theater, happens every second on Twitter, Facebook, Parler, and other social media platforms, and the proclamation and spread of disinformation and misinformation is perfectly legal. This is problematic as it can result in domestic terrorism, as seen with the 2020 Christmas Day suicide bombing in Memphis by a conspiracist opposed to 5G technology. Worse, QAnon conspiracies fueled by social media—not to mention several elected politicians publicly advocating, sharing, and legitimizing them—accelerate acts of anomie and violence by conspiracists who are QAnon adherents.[10] However, the larger danger comes from China and Russia relying on their armies of social media warriors to create and amplify these damaging conspiracies, be they about 5G, Covid-19, QAnon, election fraud, or numerous other claims, because it creates the desired schismogenesis and weakening of social cohesiveness.[11] Once this capability to attack civil society is automated, there will be a reckoning for the United States and other open societies if security institutions and strategies of resilience are not in place.

Conclusion: A New Battlespace on the Horizon?

The battlespace is new, but the challenge of adjusting to it is not. Bernard Brodie, in his 1959 *Strategy in the Missile Age*, noted that in the midst of the Cold War, as the United States and Soviet Union were developing new missile technologies, the nature of war was becoming more complex. Brodie worried that our ability to manage the demands of the future required attention to a host of threats to US security that did not fall within traditional categories. He also observed that no matter how fantastic these new technologies—what became known as the "first offset"— and their uses might seem in comparison to the past, ideas about war and how to fight it were not, even if adjustments needed to be made.

The Soviets soon matched American nuclear technology. The outcome was a balance of massive destructive power such that these adversaries could not confront one another for fear of nuclear escalation. Technological developments after this threshold of mutually assured destruction were reached and, up to the unipolar moment, occurred under the shadow of a credible nuclear second-strike capability. As nuclear weapons deterred direct confrontation, the pursuit of political ends was managed through interventions to support partners in the periphery or through war by proxy. The familiar interstitial methods of political warfare played their roles too.

There are lots of new buzzwords and calls to alarm about a completely new battlespace with its own logic and rules that make it unique. The question really comes down to whether this indirect and interstitial environment across an expanded range of domains and the associated increased importance of vulnerabilities of open societies need to be understood in a wholly different way. This book's answer is no. No, because our present reality is a continuation of the competition between states in the international system. This competition embraces the continuum of violence to meet political ends. In the present geopolitical configuration, this involves the emergence of a multipolar system and perhaps a quasi-bipolar US-China competition in the longer term.

The strategic nuclear deterrent remains an important component of this new framework. As during the Cold War, it is the ultimate arbiter in great power competition. The heart of the predicament is the challenge of articulating a clear frame for understanding the problems we face. For example, what is the consequence for our security if a national political leader knowingly gives credence to misinformation supplied by an adversarial foreign source? Interstitial warfare in the past also had this political dimension alongside the purely military problems at hand. The challenge is to take these tactical and operational problems associated with political warfare, lawfare, influence operations, and so forth across these expanded domains and nest them in a political determination of the strategic problem.

Our adversaries have been successful at harnessing new technologies to engage in new tactics to gain incremental advantage at a strategic level. Our future undoubtedly will be replete with rapid developments in information technology, artificial intelligence, and all manner of invention. It makes sense that adversaries moved to figure out the social elements of these technologies in their asymmetrical confrontations with the United States. It is not the technology, per se, that affords advantage; it is the ingenuity to harness it for advantage that counts.

Another way to think of this is to think of the United States as being in a position like Britain after its victory in World War I. Britain was a pioneer of tank technology, but Germany brought together radio, aircraft, and a novel command structure to use this technology to create a strategic effect in blitzkrieg operations. Britain and France had the same equipment and had thinkers who came up with the same organizational ideas, but they lacked the synthesis of these disparate technologies.

The preferable contemporary outcome for the United States should be to turn adversaries' strengths into weaknesses and help partners match their strengths against them. A rough list of US strengths includes alliance networks that China and Russia will be hard-pressed to match. Again, a political problem arises: Do the US political leadership and public have the interest or the political will to maintain these alliances? The United States can tap the superior track record of open societies in producing new concepts and combinations of technologies that China and Russia find hard to match. Again, the problem has a political dimension: how to break through the stultifying bureaucracy of the Department of Defense to collaborate with innovators and to modify legal frameworks to make new arrangements possible.

Notes

1. Przeworski, *Crisis of Democracy*, chap. 9.
2. Lt. Gen. Eric J. Wesley and Col. Robert H. Simpson (Ret.), "Expanding the Battlefield: An Important Fundamental of Multi-domain Operations," Association of the United States Army, April 2020.
3. Associated Press, "Putin Signs Russian Law to Shut 'Undesirable' Organizations," *Wall Street Journal*, May 23, 2015, www.wsj.com/articles/putin-signs-russian-law-to-shut-undesirable-organizations-1432404789.
4. Schelling, *The Strategy of Conflict*.
5. "Pentagon Investigated Suspected Russian Directed-Energy Attacks on U.S. Troops," *Politico*, April 22, 2021, www.politico.com/news/2021/04/22/pentagon-russia-attacks-us-troops-484150.
6. Guy, *The Weaponization of Quantum Physics*.
7. Rebecca Fannin, "China Gains to 21% of Patent Filings Globally, Right Behind US Leader at 22%," *Forbes*, April 28, 2019.
8. Francis Gurry, "China Becomes Top Filer of International Patents in 2019 amid Robust Growth for WIPO's IP Services, Treaties and Finances," World Intellectual Property Organization, April 7, 2020.
9. McMaster, *Battlegrounds*.
10. Lois Beckett, "QAnon: A Timeline of Violence Linked to the Conspiracy Theory," *The Guardian*, October 16, 2020.
11. Joseph Menn, "Russian-Backed Organizations Amplifying QAnon Conspiracy Theories, Researchers Say," *Reuters*, August 24, 2020; Julian E. Barnes, Matthew Rosenberg, and Edward Wong, "As Virus Spreads, China and Russia See Openings for Disinformation," *New York Times*, October 6, 2020; Bret Schafer, "Foreign Amplification of Voter Fraud Narratives: How Russian, Iranian, and Chinese Messengers Have Leveraged Post-election Unrest in the United States," Alliance for Securing Democracy, November 24, 2020.

8
The Future of Grand Strategy

Democracy is the most demanding of all forms of government in terms of the energy, imagination, and public spirit required of the individual.

—Gen. George C. Marshall

THINKING REALISTICALLY ABOUT CHALLENGING SITUATIONS IS ALWAYS a worthwhile pursuit. Perhaps the authors of this book are wrong, and the liberal internationalists who have dominated the American foreign policy establishment and helped shape the foreign policies of presidents Bill Clinton, George W. Bush, and Barack Obama are correct: conflict is not the default condition of the international system. Maybe those who would oppose US power can be convinced that it would be better for them if the underlying logic of strategic thinking and the nature of warfare no longer applied.

One of the main points of this book is that we have been here before. Winston Churchill wrote in his history of World War I, *The World Crisis*, of such optimism surrounding Britain's role in resolving the Crisis of Agadir in 1911:

> Soft, quiet voices purring, courteous, grave, exactly-measured phrases in large peaceful rooms. But with less warning cannons had opened fire and nations had been struck down by this same Germany. So now the Admiralty wireless whispers through the ether to the tall masts of ships, and captains pace their decks absorbed in thought. It is nothing.

It is less than nothing. It is too foolish, too fantastic to be thought of in the twentieth century. Or is it fire and murder leaping out of the darkness at our throats, torpedoes ripping the bellies of half-awakened ships, a sunrise on a vanished naval supremacy, and an island well guarded hitherto, at last defenceless? No, it is nothing. No one would do such things. Civilisation has climbed above such perils. The interdependence of nations in trade and traffic, the sense of public law, the Hague Convention, Liberal principles, the Labour Party, high finance, Christian charity, common sense have rendered such nightmares impossible. Are you quite sure? It would be a pity to be wrong. Such a mistake could only be made once—once for all.[1]

In 1940 E. H. Carr expressed similar skepticism of what he called the "modern school of utopian thought" in his *The Twenty Years' Crisis, 1919–1939*, arguing, "The exposure by realist criticism of the utopian edifice is the most urgent task of the moment in international thought."[2] The overall record appears to be fairly consistent.

This is not to say that World War III is right around the corner. This is the juncture at which the reader is reminded of the GRINS framework used here to show how state behavior is based on geopolitical contexts, the nature of regimes, ideas, types of military organization, and the development of technology. How wars are fought changes a great deal, even as why wars are fought remains remarkably constant. Interstitial warfare is easily anticipated in circumstances in which weaker states challenge stronger ones and between strong states that could annihilate each other in all-out war. Balancing power relationships in the international system is a constant too, and the US currently engages in this approach with other countries to contain China's rise.

Another purpose of this book has been to use history in a comparative fashion to assist in thinking about how to address the significant changes in the character of warfare, especially the expansion of domains of contention—most particularly the intensified accessibility to influence individuals in open societies that adversaries enjoy—and the diffusion of new technologies in ways that diminish US dominance in weaponry.

The Contours of the Challenge

The United States is facing a crisis of grand strategy and military strategy. The challenge of military strategy mirrors, and is a function of failures in, US grand strategy. Grand strategy aims to harness all instruments of national power (i.e., diplomatic, informational/ideational, military, economic) toward realizing a specific national objective. Military strategy, a subset of grand strategy, reconciles military ends, ways, and means in pursuit of a military objective that aligns with the broader national objective. In the absence of a coherent grand strategy or political end state, military strategy necessarily remains adrift if not misdirected. After decades of continuous conflict, the fact that the United States is unable to turn its tactical gains into strategic advantages is an illustration of how the disconnect between grand strategy and military strategy turns military efforts into an exercise in futility. More recently, the United States played the defining role in defeating the Islamic State in Iraq and Syria (ISIS). However, not nesting the fight against ISIS in a broader strategic framework meant that tactical expediency replaced strategy, and though the United States won that battle, its adversaries have gained the strategic advantage at America's expense. America will not be able to afford such strategic profligacy as it enters an era of great power contestation.

The United States encounters this strategic crisis just as it enters a transformed global environment. It and its allies can no longer claim military, economic, and ideational hegemony. Nondemocratic adversaries are empowered, and emboldened and revisionist powers are explicit about their aspirations. Inevitably, the United States will be forced to formulate a grand strategy that is consistent with its values to address challenges of the present and the future. In that moment, the United States (and its allies) will be forced to reckon with the institutional challenges side by side with the external and internal origins of the current strategic crisis.

Though complex, the contemporary geopolitical context retains a series of legacies that provide a sense of continuity. Nuclear deterrence remains in place, though it faces the challenge of nuclear proliferation. Is it too risky to engage in any large-scale military option that is likely to involve a nuclear-armed state? Nine states possess nuclear weapons. The United States and Russia each have about 4,000 operative weapons, while China has over 300 weapons. France, the United Kingdom, Pakistan, India, Israel, and North Korea each have enough nuclear weaponry to inflict catastrophic damage. Does this situation, as well as the probable expansion of this list of countries, mean that the old logic of successful deterrence extends to all, or will the odds of miscalculation and poor signaling mean eventual catastrophe in at least one region of the world, with considerable global geopolitical consequences?

This persistence of nuclear weapons and the development of devastating nonnuclear weapons probably means that warfare will continue its trend since 1945 of becoming resolutely interstitial. That is, limited warfare in all its (expanding) dimensions might be the way of the future. Deterrence is much less straightforward in this environment, and communication through the use of force, which Thomas Schelling saw as so central to avoiding nuclear war in the middle of the twentieth century, is more difficult to manage in the expanded battlespace in which interstitial challenges now occur. But thinking about deterrence is useful in addressing challenges of this sort. For example, what does the use of force convey to the adversary, say, in a cyber domain, and can it have any deterrent value within that domain or by extension in other domains? Does deterrence in these domains require the maintenance of conventional military forces to signal the resolve to fight?

In this context, US failures in military strategy reflect failures in grand strategy, which is where military and political problems meet. Though the focus here is on the challenges of military strategy, given the inseparability between grand and military strategy, a brief outline of the crisis in grand strategy precedes the discussion of military strategy. This crisis in grand strategy has *ideational* and *political* origins. The challenges of military strategy necessarily reflect challenges of grand strategy side by side with military-specific *institutional*, *material*, *doctrinal*, and *personnel* origins.

US grand strategy put in place to win the Cold War—best framed in National Security Council Paper 68—worked. At the end of the Cold War, the United States and its allies emerged with a preponderance of (collective) economic and ideational capital. The institutional innovations and material, doctrinal, and personnel changes put in place through the Second Offset Strategy paid dividends at the end of the Cold War by making the United States the unrivaled military power, turning the world militarily into a unipolar world. Specifically, at the *institutional* level, the Goldwater-Nichols Department of Defense Reorganization Act operationalized the Joint Force concept by instituting sweeping changes across branches. Services retained the authority to organize, train, and equip forces and make them combat ready for use by the combatant commands. The combatant commands were given the authority to conduct operations across the spectrum of conflict once so directed by the president or the secretary of defense. The Unified Command Plan, modified every few years, delineates the areas and domains of responsibility for each of the geographic and functional combatant commands. Functional combatant commands operate worldwide across geographic boundaries and provide unique capabilities to geographic combatant commands. Each combatant commander is assigned component commanders—land, air, maritime, and special operations. Making these component commanders report to the unified commander rather than service chiefs facilitates unity of command and minimizes interservice rivalry.[3]

This institutional-level solution was complemented with *material* solutions. Given the inability of the North Atlantic Treaty Organization (NATO) to quantitatively match the numbers of Warsaw Pact military forces, the United States looked to technology to serve as a force multiplier, to qualitatively exceed the Warsaw Pact. If it could not balance military power soldier for soldier or tank for tank, the United States could leverage technology to multiply the value of a US tank, soldier, airplane, and so forth, relative to its Warsaw Pact counterpart. The United States labeled this the "Offset Strategy," now typically referred to as the "Second Offset Strategy," with the first being nuclear weapons.[4] Until this day, US combat primacy is predicated on developments in the second offset. "The Second Offset Strategy focused on developing new technologies in five key areas: 1) new [intelligence, surveillance, and reconnaissance (ISR)] platforms; 2) battle management capabilities (communications and C2 [Command and Control] information systems); 3) improved precision-strike weapons; 4) stealth technology aircraft; and, 5) the tactical exploitation of space for ISR, communications, and [positioning, navigation, and timing]."[5]

These technologies focused on "seeing deep" and "shooting deep," or the ability to identify and destroy targets deep behind the line of contact. Armed with the ability to see and shoot deep, US forces would attrite and disintegrate Warsaw Pact forces in depth, creating an advantageous relative force ratio prior to direct fire engagements. Ultimately, second offset innovations included short- and long-range precision-guided weapons (including the cruise missile, laser-guided bombs, and the Army Tactical Missile System), manned ISR platforms (such as the Joint Surveillance Target and Attack Radar System), armed and unarmed unmanned aerial vehicles, satellite navigation, space-based reconnaissance, and stealth technology. Simulations indicated that these technologies had the potential to revolutionize warfare.

Doctrinal solutions came in the form of developing new joint doctrine and joint tactics, techniques, and procedures—that is, from the tactical level up, service members were required to train and operate in a joint environment, becoming familiar with service-specific capabilities and cultures. *Personnel* solutions included the requirement to complete a joint assignment mandatory for promotion to flag rank and the introduction of Joint Professional Military Education institutions.

The combination of the above solutions worked to align the tactical, operational, and strategic levels under the aegis of the Joint Force concept. Operation Just Cause in Panama (1989) was the first test, followed by the Gulf War (1991), where the Joint Force performed beyond all expectations. At the end of the Cold War, the United States emerged with a global security architecture of over 800 bases worldwide and a global property portfolio valued at over $1 trillion. This military primacy was an outcome of deliberate decisions. It would be taken for granted in the subsequent years, a case of the success of the past sabotaging needed responses to the present.

Grand strategy formulations that followed the end of the Cold War took on a transformative character—an ideational enterprise—guided by (or now referred to as) liberal internationalism. There were good reasons. For example, European history until the end of the Cold War was replete with continuous warfare. The European Union that emerged under the NATO and US security umbrella was a liberal-internationalist solution that fundamentally transformed the power politics of a continent previously locked in a cycle of devastating conflict. NATO, in the words of its first secretary-general, Lord Hastings Lionel Ismay, was created to "keep the Soviet Union out, the Americans in, and the Germans down."[6] That was a very successful formula. If liberal solutions could be found to mitigate problems of power politics in Europe, could they not work everywhere?

There were two errors in turning this ideational enterprise into a grand strategy—into strategic narcissism. The first error was a fundamental misreading of geopolitical power relations. The second error was a fundamental misreading of the wellsprings of grand strategy in a liberal democracy. This is an assessment not of the merits of liberal internationalism but of its strategic formulation.

Exporting democracy and capitalism and imposing humanly devised geopolitical entities is a transformative enterprise. Any transformative enterprise generates distributional consequences with winners and losers and inevitable resistance; therefore, it is necessary to approach the world the way it is, not the way it should be. The first error in this formulation of grand strategy was in conflating the ultimate desired strategic end state—what the world should be—with the way the world actually is; that is, the operational environment in which the strategy was to be implemented (the way the world is) was interpreted in terms of the desired end state (the way the world should be). When operationalized, this disconnect—a strategic bait and switch, a theoretical conflation devised for purposes of political expediency—generates incongruities that adversaries can exploit. Extension of this logic militarily is open-ended commitments with no discernible strategic gains, and an extension of this logic economically is access to American markets with no reciprocity, as was the case during the Cold War.

The second error was a misreading of the ultimate source of power in any grand strategy in a democracy. No strategy is sustainable in a democracy without buy-in from citizens. A social contract between the people and grand strategy is a necessary condition because the people will want to be convinced as to why their loved ones should be sent into harm's way and why they should bear the burden of foreign policy decisions. During the Cold War, policymakers actively attempted to rein in selective business interests knowing that American global leadership requires a strong economy and a content citizenry at home. At the end of the Cold War, liberal internationalists who had come of age during the Cold War failed to revisit the social contract between the people and the government that underpinned Western grand strategy during the Cold War. They took it for granted. By taking the American social contract—which generates an ideational consensus—for granted, subsequent administrations engaged in strategic profligacy.

In this strategic crisis, the primary claim is not about making a theoretical point on the merits of the liberal-internationalist vision of what the world should be. And it is certainly not about whether the world would indeed be a better place if it were governed by transnational juridical entities and institutions. One would not know, for that is not the way the world is at the moment. By acting as if the world were the way it should be, by turning acting-as-if into America's grand strategy, and by not forcing others to do the same and letting them get away with no repercussions, the US failed comprehensively in its strategic execution. The contemporary crisis in military strategy stems directly from this failure of grand strategy.

Challenges of Contemporary Military Strategy in the Multidomain Environment

At the strategic level, the increasing use of the US military with no clearly discernible political end states has afflicted the American armed forces with Groundhog Day Disease. In this situation the military force is tactically and operationally very proficient, but all of that proficiency is wasted when the country's political leadership is unable to articulate a viable strategic vision. The US strategic failure has a similar impact in the misapplication of force and the devaluing of tactical and operational successes.

For example, at the end of the Gulf War in 1991, the United States committed to maintaining a no-fly zone over Iraq (at the cost of over $1 billion per year). The United States also engaged in the Balkans while maintaining the no-fly zone and continuing to uphold US commitments in the Pacific. This was another example of how the United States' investment in its military during the Cold War continued to pay dividends. While US commitments in the Pacific remained unchanged, the expansion of NATO extended US military commitments. However, the commitment had clearly delineated end states. With the onset of the Global War on Terror, the United States entered an era of open-ended warfare, mainly in the Middle East but also in the Pacific and sub-Saharan Africa, that has continued for two decades. Open-ended commitments with no political end states lead the military to continue engaging militarily, whereby application of military power without a political strategy becomes its own enterprise even though it is a useless exercise—resulting in Groundhog Day Disease.

Alongside the challenges of military strategy in a context in which there appear to be no clearly articulated political end states to align military end states, armed forces are also facing their own institutional, material, doctrinal, and personnel challenges, which have strategic ramifications in the present operational environment.

US armed forces are entering a multidomain operational environment. In the American conception, the domains include land, sea, air, cyber, and space. The objective in a highly integrated multidomain environment is to leverage the advantages in one domain to gain an advantage in another so that the United States can maintain varying degrees of advantage in all domains. In this operational environment, adversaries of the United States have an advantage as they operate from an ethical, moral, legal, and political paradigm at variance from that of the United States. Therefore, the United States faces challenges of military strategy on the *doctrinal, institutional, material,* and *personnel* levels.

The doctrinal challenge is in accurately conceptualizing the adversarial paradigm and outflanking adversaries' conceptual flanks without compromising democratic principles. That may mean the United States will have to carry a greater burden, but in doing so, it can gain space and depth vis-à-vis its adversaries. In sum, the United States will have to frame an offset strategy, where it is able to fight, win, and dominate across the full spectrum of the multidomain operational environment.

Doctrinal innovations must accompany an evaluation of current US defense institutional architecture at the strategic level. The Joint Force structure and the accompanying Goldwater-Nichols Department of Defense Reorganization Act enabled the United States to dominate the military dimension in the Cold War and helped it remain the military hegemon for three decades afterward. The Joint Force and combatant command concept certainly alleviated interservice rivalry and helped the United States coalesce military power across domains. However, is the Joint Force architecture, with its administrative and operational chains of commands and geographic and functional combatant commands, the best institutional configuration in a multidomain operational environment? In a global strategic contest, geographic and functional commands transcended service rivalries, but it would not be an exaggeration to say that interservice rivalries are supplanted by combatant command rivalries for resources. In a multidomain operational environment, where geographic boundaries and areas of responsibility become less salient, institutional innovations that seek to better integrate command and control across commands will be necessary. At the operational level, far more than geographic commands, functional commands will have to evaluate the ways that they can assist in the multidomain fight—precisely because America's functional and geographic commands can no longer take for granted that they can continue to engage in uncontested operational environments.

Concurrent doctrinal and institutional innovation will necessarily entail a discussion on the strategic use of covert assets under Title 50 (intelligence agency authority) and Title 10 (military authority) in support of Title 50. As the nuclear deterrence probably will hold, adversaries will always fight below the threshold of war—engaging in interstitial tactics. It will be incumbent upon the United States to evaluate and then expand its Title 50 assets where they can in turn be deployed in pursuit of long-term strategic objectives. It might not surprise the reader that the rapid expansion of Title 10 (military-run covert and clandestine operations) causes friction with what is now a proportionally smaller Title 50 (CIA operations). The secretary of defense can have good reason and has the legal power to assign Title 50 operations to the military.

Material and personnel challenges walk in lockstep. In land, air, sea, and space, the United States is about to expend the last of the material advantages left from its investment in the Second Offset Strategy. Though each branch has innovated at the tactical level in the last twenty years, the dominant assets on land, sea, and air, as well as the nuclear triad, are the last vestiges of the second offset that each service will keep refurbishing and reconditioning, even if inefficiently, as noted earlier in the remarkable survival of the very expensive F-35 fighter aircraft. It serves the purpose, in uncontested air, land, and maritime domains, of fighting adversaries with great asymmetries in capabilities, but these stopgap measures are hardly adequate when, as is bound to happen soon, the United States is confronted with the reality of operating in contested air, sea, land, space, and cyber domains.

The material challenge with strategic ramifications also mirrors, and is a function of failures in, grand strategy. Open-ended grand strategy took military power for granted, and there were no overall assessments of how best to reconcile ways, means, and ends at the strategic level for today and the future. A historical example in the air domain sheds light on how material-level challenges with strategic ramifications mirror strategic ineptitude. During the Vietnam War, the United States did not come anywhere near air superiority, let alone air supremacy. After all, the pilot and future US senator and presidential candidate John McCain was shot down over North Vietnam in 1967 and remained a prisoner until 1973. Sam Johnson, the representative to the US Congress from Texas's 3rd District, spent nearly seven years as a prisoner of war after he was shot down over North Vietnam. Once in Congress, these two proved to be keen promoters and defenders of air superiority! US pilots after the Cold War faced nothing approaching the risks of enemy fire that Vietnam War–era pilots faced.

Air superiority means being able to conduct air operations without prohibitive interference by the opposing force; air supremacy is when the opposing air force is incapable of effective interference, and one can completely dominate the air space. The United States implemented the second offset in the post-Vietnam era, when it was assessing its failures and reevaluating its military strategy in terms of the broader grand strategy to gain the military advantage vis-à-vis the Soviet Union. US material solutions put in place still dominate our most relied-upon air assets, from America's strategic bomber fleet, fighters, and ISR assets to command-and-control assets that allowed the US Air Force to exercise its primary missions of global strike, air superiority, and global mobility (to name three out of the six) seamlessly. While the United States has developed niche assets that are no longer unrivaled (such as the F-35, the air version of the Bradley Fighting Vehicle, a weapons system designed by committee that has taken on a life of its own), no strategic-level evaluation has taken place on the whole of the national level to come up with the sort of material solutions the United States would need to face and dominate the multidomain operational environment.

The Future of Grand Strategy 169

Personnel-level challenges also mirror the lack of strategic direction to ready the force for a future fight. The United States provides for soldiers in terms of recruitment, pay, rank, and responsibilities as if it is readying to fight another war of industrial organizations. Personnel challenges—since the French mass conscription of Napoleon's time to the end of the Cold War in France and end of Vietnam in the United States—were always seen as a thorny problem with an easy solution: the compulsory draft. But as Vietnam showed for the Americans, open-ended interstitial wars with vague and changing endpoints draw a great deal less public ire when fought by professional armed forces.

Though there will always be a place for grunts and quartermasters and their service will always be invaluable, the United States is entering the graduate level of warfare in a multidomain environment. In the air force and to a lesser degree in the navy, officers with degrees are the trigger pullers and war fighters. But in the army and the marines, it is the enlisted personnel who are the war fighters, and their average age remains eighteen to twenty-two; a majority have no college degree, and only a limited number of noncommissioned officers have college degrees. It's not that a college degree will automatically equip individuals to effectively engage in a multidomain operational environment, only that the nature of US recruitment strategies reflects the atrophied nature of military thinking when it comes to readying the force and building the team to dominate in the multidomain fight. Military service already remains a nearly esoteric enterprise in the broader society where under 1 percent of the population serves and, for over 80 percent of that 1 percent, service is a family tradition. Readying the force for a future fight will require evaluating existing assumptions regarding training, remuneration, and recruitment, side by side with a broader discussion of civic obligations and the duties of citizenship in a liberal democracy.

Connecting the Political and the Military in a Future Strategy

Addressing challenges in military terms is necessary but not sufficient. It is a core point of this book that successful strategy rests upon describing how military means will achieve specific ends. The military element, which must adjust to specific circumstances described in this book, involves the direction and use of force and the threat of force to achieve these ends. These ends are policies that are decided by politics, and this is where a large part of the challenge lies. As described earlier, the character of US politics and of open societies places the United States at a disadvantage in some respects vis-à-vis authoritarian states. This is not to say that the United States should adopt authoritarian features to meet challenges from authoritarian states. There are, however, some points for political reflection in devising a coherent strategy in the event that bad times return.

Some of these political considerations include the following:

• What are the limits of speech and the dissemination of information? Social media companies now regulate "uncivil speech." The problem is more serious when public figures make false statements, refer to nonexistent events, and promote conspiracy theories. The divisive nature of this speech, intended to advance domestic political prospects, presents an ideal attack surface for adversaries that coordinate with domestic political figures, regardless of these individuals' preferences or level of knowledge about this process.

• How does a foreign policy establishment emerge from a period of "strategic narcissism"? Foreign policy experts have been criticized for elitism since the end of World War II. That is a misrepresentation of an acute problem. The complexity of foreign policy requires relying on subject matter experts. They are castigated as elites not because they are elites but because they have become an exclusivist epistemic community and are never held accountable. They can be consistently wrong yet manage to recycle themselves into each new administration. The strategic ramification of this exclusive epistemic community is that they become the conceptual and doctrinal gatekeepers. These are gatekeepers who refuse to question their assumptions and frames of references and balk at shifting the foreign policy paradigm no matter how wrong—for their paradigmatic positions have become articles of faith, not reason. Removing these obstacles to change overtime will be imperative if one is to shift the United States into a new strategic paradigm.

• In that moment . . . armed forces will need to craft a military strategy—with institutional-, doctrinal-, material-, and personnel-level innovations in mind.

• US political leaders will also need to craft a new social contract in support of the grand strategy. Here's hoping . . .

In any event, those engaged with policy and politics will benefit from how insights from the Napoleonic era through the recently concluded unipolar moment are relevant to the theory and practice of contemporary strategy. A serious consideration of strategy rooted in historical experience is even more important in our era in which considerable attention is focused on the emergence of the novel and the seemingly novel. New adversaries, new domains of warfare, and novel weapons are emerging, but carefully delineating what is genuinely different from what is timeless is our way of offering guidance.

Notes

1. Churchill, *The World Crisis*, 1:45, https://babel.hathitrust.org/cgi/pt?id=uc1.32106006367657&view=1up&seq=9.
2. Carr, *The Twenty Years' Crisis, 1919–1939*, 113.
3. Brent J. Talbot, "Goldwater-Nichols and the Evolution of the Joint Force," in Burke, Fowler, and McCaskey, *Military Strategy, Joint Operations, and Airpower*.
4. Col. Hugh W. A. Jones, *Multi-domain Operations—Expanding the Battlefield and Re-establishing America's Military Offset* (Army War College Fellows, Strategy Research Project, March 2021), 62.
5. Jones, *Multi-domain Operations*, 64.
6. "Lord Ismay," NATO, www.nato.int/cps/us/natohq/declassified_137930.htm.

Acronyms

AI	Artificial intelligence
CCP	Chinese Communist Party
CJCS	Chairman of the Joint Chiefs of Staff
DIME	Diplomatic, informational/ideational, military, economic
EU	European Union
GDPR	General Data Protection Regulation (European Union)
GPS	Global Positioning System
GRINS	Geopolitics, regime type, ideas, nature of military organizations, scientific knowledge
ICC	International Criminal Court
IGO	Intergovernmental organization
IRGC	Iranian Revolutionary Guard Corps
ISIS	Islamic State in Iraq and Syria
ISR	Intelligence, surveillance, reconnaissance
JCPOA	Joint Comprehensive Plan of Action
KGB	Committee for State Security (Komitet Gosudarstvennoy Bezopasnosti)
MISO	Military information support operations
NATO	North Atlantic Treaty Organization
	Nongovernmental organization
NRA	National Rifle Association
R2P	Responsibility to protect
UBL	Usama Bin Laden

UN	United Nations
UNOSOM	United Nations Operation in Somalia
USAID	United States Agency for International Development
USIA	United States Information Agency
WMDs	Weapons of mass destruction

Bibliography

Acemoglu, Daron, and James A. Robinson. *Why Nations Fail: The Origins of Power, Prosperity, and Poverty.* New York: Random House, 2012.
Acheson, Dean. *Present at the Creation: My Years in the State Department.* New York: W. W. Norton & Company, 1970.
Adams, Jefferson. *Strategic Intelligence in the Cold War and Beyond.* New York: Routledge, 2013.
Allen, T. S., and A. J. Moore. "Victory Without Casualties: Russia's Information Operations." *Parameters* 48, no. 1 (2018): 59–71.
Ambrose, Stephen E. *Eisenhower: Soldier and President.* New York: Simon & Schuster, 1991.
Anderson, Benedict. *Imagined Communities: Reflections on the Origin and Spread of Nationalism.* New York: Verso, 2006 [1983].
Andress, David. *The Terror: Civil War in the French Revolution.* London: Little, Brown, 2005.
Andrew, Christopher, and Vasili Mitrokhin. *The World Was Going Our Way: The KGB and the Battle for the Third World.* New York: Basic Books, 2005.
Applebaum, Anne. *Twilight of Democracy: The Seductive Lure of Authoritarianism.* New York: Doubleday, 2020.
Armitage, David. *Civil Wars: A History in Ideas.* New Haven, CT: Yale University Press, 2017.
Arnold, Thomas D., and Nicolas Fiore. "Five Operational Lessons from the Battle for Mosul." *Military Review* (January–February 2019): 56–71.
Åslund, Anders. *Russia's Crony Capitalism: The Path from Market Economy to Kleptocracy.* New Haven, CT: Yale University Press, 2019.
Bail, Christopher A., et al. "Exposure to Opposing Views on Social Media Can Increase Political Polarization." *Proceedings of the National Academy of Sciences* 115, no. 37 (2018): 9216–9221.
Bailey, Andrew, et al., eds. *The Broadview Anthology of Social and Political Thought,* Vol. 2: *The Twentieth Century and Beyond.* Buffalo, NY: Broadview Press, 2008.

Barfield, Thomas. *Afghanistan: A Cultural and Political History*. Princeton, NJ: Princeton University Press, 2010.

Barma, Naazneen H. "Peace-Building and the Predatory Political Economy of Insecurity: Evidence from Cambodia, East Timor and Afghanistan," *Conflict, Security & Development* 12, no. 3 (2012): 273–298.

Barkey, Karen. *Bandits and Bureaucrats: The Ottoman Route to State Centralization*. Ithaca, NY: Cornell University Press, 1994.

Bateson, Gregory. "Culture Contact and Schismogenesis." *Man* 35 (December 1935): 178–183.

Beckley, Michael. "Economic Development and Military Effectiveness." *Journal of Strategic Studies* 33, no. 1 (2010): 43–79.

———. "The Power of Nations: Measuring What Matters." *International Security* 43, no. 2 (2018): 7–44.

Ben-Ari, Eyal, et al. *Rethinking Contemporary Warfare: A Sociological View of the Al-Aqsa Intifada*. Albany: State University of New York Press, 2010.

Benkler, Yochai, Robert Faris, and Hal Roberts. *Network Propaganda: Manipulation, Disinformation, and Radicalization in American Politics*. New York: Oxford University Press, 2018.

Berdal, Mats R., and David Malone, eds. *Greed and Grievance: Economic Agendas in Civil Wars*. Boulder, CO: Lynne Rienner Publishers, 2000.

Bergsmo, Morten, and Emiliano J. Buis, eds. *Philosophical Foundations of International Criminal Law: Correlating Thinkers*. Brussels: Torkel Opsahl Academic EPublisher, 2018.

Berlin, Isaiah. *Against the Current: Essays in the History of Ideas*. Princeton, NJ: Princeton University Press, 2013 [1955].

———. *Four Essays on Liberty*. Oxford: Oxford University Press, 1969.

———. *Two Concepts of Liberty*. London: Oxford University Press, 1977 [1958].

Berman, Eli, Joseph H. Felter, and Jacob N. Shapiro. *Small Wars, Big Data: The Information Revolution in Modern Conflict*. Princeton, NJ: Princeton University Press, 2018.

Berman, Sheri. "Civil Society and the Collapse of the Weimar Republic." *World Politics* 49, no. 3 (1997): 401–429.

Biddle, Stephen D. *Military Power: Explaining Victory and Defeat in Modern Battle*. Princeton, NJ: Princeton University Press, 2004.

Bøås, Morten. "Liberia and Sierra Leone—Dead Ringers? The Logic of Neopatrimonial Rule." *Third World Quarterly* 22, no. 5 (2001): 697–723.

Bogaards, Matthijs. "Iraq's Constitution of 2005: The Case Against Consociationalism 'Light.'" *Ethnopolitics* (August 23, 2019): 1–17.

Brands, Hal. *Latin America's Cold War*. Cambridge, MA: Harvard University Press, 2010.

Brecht, Richard, and William Rivers. *Language and National Security: The Federal Role in Building Language Capacity in the U.S.* Washington, DC: Department of Education, 2001.

Breuilly, John. *Austria, Prussia and the Making of Germany: 1806–1871*. New York: Routledge, 2014.

Brodie, Bernard. *Strategy in the Missile Age*. Princeton, NJ: Princeton University Press, 1957.

Broniatowski, David A., et al. "Weaponized Health Communication: Twitter Bots and Russian Trolls Amplify the Vaccine Debate." *American Journal of Public Health* 108, no. 10 (2018): 1378–1384.

Brooks, Risa A., and Elizabeth A. Stanley, eds. *Creating Military Power: The Sources of Military Effectiveness*. Palo Alto, CA: Stanford University Press, 2007.

Brooks, Rosa. *How Everything Became War and the Military Became Everything: Tales from the Pentagon*. New York: Simon & Schuster, 2016.

Brose, Christian. *The Kill Chain: Defending America in the Future of High-Tech Warfare*. New York: Hachette Books, 2020.

Brownlee, Billie Jeanne. *New Media and Revolution: Resistance and Dissent in Pre-uprising Syria*. Montreal: McGill-Queen's Press, 2020.

Bruton, Bronwyn. *Twenty Years of Collapse and Counting: The Cost of Failure in Somalia*. Washington, DC: Center for American Progress, 2011.

Brzezinski, Zbigniew. *The Grand Chessboard: American Primacy and Its Geostrategic Imperatives*. New York: Basic Books, 2016 [1997].

Buchanan, Ben. *The Hacker and the State: Cyber Attacks and the New Normal of Geopolitics*. Cambridge, MA: Harvard University Press, 2020.

Bunker, Robert J. "Unconventional Warfare Philosophers." *Small Wars & Insurgencies* 10, no. 3 (1999): 136–149.

Burke, Ryan, Michael Fowler, and Kevin McCaskey. *Military Strategy, Joint Operations, and Airpower: An Introduction*. Washington, DC: Georgetown University Press, 2018.

Burke, Ryan, and Jahara Matisek. "The Illogical Logic of American Entanglement in the Middle East." *Journal of Strategic Security* 13, no. 1 (2020): 1–25.

Butt, Ahsan I. "Why Did the United States Invade Iraq in 2003?" *Security Studies* 28, no. 2 (2019): 250–285.

Cardwell, Curt. *NSC 68 and the Political Economy of the Early Cold War*. New York: Cambridge University Press, 2011.

Carr, E. H. *The Twenty Years' Crisis, 1919–1939: An Introduction to the Study of International Relations*. London: Macmillan, 1962 [1939].

Carson, Austin. *Secret Wars: Covert Conflict in International Politics*. Princeton, NJ: Princeton University Press, 2018.

Caspersen, Nina. *Contested Nationalism: Serb Elite Rivalry in Croatia and Bosnia in the 1990s*. New York: Berghahn Books, 2010.

Chen, Dan. "Political Context and Citizen Information: Propaganda Effects in China." *International Journal of Public Opinion Research* 31, no. 3 (2019): 463–484.

Chivers, Christopher John. *The Gun*. New York: Simon & Schuster, 2011.

Churchill, Winston. *The World Crisis, 1911–1914*, Vol. 1. New York: Charles Scribner's Sons, 1924.

Clarke, Richard A., and Robert K. Knake. *The Fifth Domain: Defending Our Country, Our Companies, and Ourselves in the Age of Cyber Threats*. New York: Penguin Press, 2019.

Clausewitz, Carl von. *On War*. Translated and edited by Michael Howard and Peter Paret. Princeton, NJ: Princeton University Press, 1989 [1832].

Clinton, Hillary Rodham. *Hard Choices: A Memoir*. New York: Simon & Schuster, 2014.

Cooley, Alexander, and Daniel Nexon. *Exit from Hegemony: The Unraveling of the American Global Order*. New York: Oxford University Press, 2020.

Craig, Dylan. "Intermediarized Security Governance and the 'Sultans' Retort.'" *Journal of the Middle East and Africa* 5, no. 2 (2014): 131–149.

Crockatt, Richard. *The Fifty Years War: The United States and the Soviet Union in World Politics, 1941–1991*. New York: Routledge, 1995.

Czosseck, Christian, and Kenneth Geers, eds. *The Virtual Battlefield: Perspectives on Cyber Warfare*. Amsterdam: IOS Press, 2009.

Davidson, James Dale, and Lord William Rees-Mogg. *The Sovereign Individual: Mastering the Transition to the Information Age*. New York: Simon & Schuster, 2020 [1997].

Donnithorne, Jeff. *Four Guardians: A Principled Agent View of American Civil-Military Relations*. Baltimore, MD: Johns Hopkins University Press, 2018.
Douhet, Giulio. *The Command of the Air*. Translated by Dino Ferrari. Washington, DC: Air Force History and Museum Project, 1998.
Downing, Brian. *The Military Revolution and Political Change: Origins of Democracy and Autocracy in Early Modern Europe*. Princeton, NJ: Princeton University Press, 2020.
Dunlap, Maj. Gen. Charles. "Lawfare 101: A Primer." *Military Review* (May/June 2017): 8–17.
Durham, Norman L. *The Command and Control of the Grand Armée: Napoleon as Organizational Designer*. Auckland: Pickle Partners Publishing, 2015.
Durkheim, Émile. *The Division of Labour in Society*. Translated by W. D. Halls. New York: Free Press, 1997 [1893].
Duyvesteyn, Isabelle, and Jan Angstrom, eds. *Rethinking the Nature of War*. New York: Frank Cass, 2005.
Dwyer, Philip G. *Modern Prussian History: 1830–1947*. New York: Routledge, 2014.
Edwards, Jason A. "Make America Great Again: Donald Trump and Redefining the US Role in the World." *Communication Quarterly* 66, no. 2 (2018): 176–195.
Farrell, Henry, and Abraham L. Newman. "Weaponized Interdependence: How Global Economic Networks Shape State Coercion." *International Security* 44, no. 1 (2019): 42–79.
Fearon, James D. "Rationalist Explanations for War." *International Organization* 49, no. 3 (1995): 379–414.
Fearon, James D., and David D. Laitin. "Ethnicity, Insurgency, and Civil War." *American Political Science Review* 97, no. 1 (2003): 75–90.
Febvre, Lucien, and Henri-Jean Martin. *The Coming of the Book: The Impact of Printing, 1450–1800*. New York: Verso, 1997.
Ferguson, R. Brian. "Masculinity and War." *Current Anthropology* 62, no. S23 (2021): S112–S124.
Fink, Christina. "Dangerous Speech, Anti-Muslim Violence, and Facebook in Myanmar." *Journal of International Affairs* 71, no. 1.5 (2018): 43–52.
Fiott, Daniel. "A Revolution Too Far? US Defence Innovation, Europe and NATO's Military-Technological Gap." *Journal of Strategic Studies* 40, no. 3 (2017): 417–437.
Fraser, George MacDonald. *The Flashman Papers: The Complete 12-Book Collection*. New York: Harper Collins, 2013 [1969–2005].
Freedman, Lawrence. *The Future of War: A History*. New York: PublicAffairs, 2017.
Friedberg, Aaron L. "Why Didn't the United States Become a Garrison State?" *International Security* 16, no. 4 (1992): 109–142.
Friedman, Thomas. *The World Is Flat: A Brief History of the Twenty-First Century*. New York: Farrar, Straus and Giroux, 2005.
Fukuyama, Francis. "The End of History?" *National Interest*, no. 16 (summer 1989): 3–18.
———. *The End of History and the Last Man*. New York: Simon & Schuster, 2006 [1992].
Gaddis, John Lewis. *George F. Kennan: An American Life*. New York: Penguin Books, 2012.
Galeotti, Mark. "Hybrid, Ambiguous, and Non-linear? How New Is Russia's 'New Way of War'?" *Small Wars & Insurgencies* 27, no. 2 (2016): 282–301.
Gall, Carlotta. *The Wrong Enemy: America in Afghanistan, 2001–2014*. Boston: Houghton Mifflin Harcourt, 2014.

Gamson, William A., et al. "Media Images and the Social Construction of Reality." *Annual Review of Sociology* 18, no. 1 (1992): 373–393.
Gartzke, Erik. "War Is in the Error Term." *International Organization* 53, no. 3 (1999): 567–587.
Gentry, Curt J. *Edgar Hoover: The Man and the Secrets*. New York: W. W. Norton & Company, 2001.
Gerasimov, Valery Vasilevich. "New Challenges Demand Rethinking Forms and Methods of Combat Action." *Industrial Courier*, February 27, 2013.
Gevorkyan, Nataliya, Natalya Timakova, and Andrei Kolesnikov. *First Person: An Astonishingly Frank Self-Portrait by Russia's President Vladimir Putin*. New York: PublicAffairs, 2000.
Gibbs, Montgomery B. *Military Career of Napoleon the Great: An Account of the Remarkable Campaigns of the "Man of Destiny."* Chicago: Werner Company, 1895.
Gibson, Christopher Patrick. *Securing the State: Reforming the National Security Decisionmaking Process at the Civil-Military Nexus*. Farnham, UK: Ashgate Publishing, 2008.
Glenn, John K. *Framing Democracy: Civil Society and Civic Movements in Eastern Europe*. Palo Alto, CA: Stanford University Press, 2003.
Goertzel, Ted. "Belief in Conspiracy Theories," *Political Psychology* (1994): 731–742.
Goertzel, Ted. *Turncoats and True Believers: The Dynamics of Political Belief and Disillusionment*. Buffalo, NY: Prometheus Books, 1992.
Gorbachevsky, Boris. *Through the Maelstrom: A Red Army Soldier's War on the Eastern Front, 1942–1945*. Edited and translated by S. Britten. Lawrence: University Press of Kansas, 2008.
Gray, Colin. *The Strategy Bridge: Theory for Practice*. New York: Oxford University Press, 2010.
Green, Brendan Rittenhouse. *The Revolution That Failed: Nuclear Competition, Arms Control, and the Cold War*. New York: Cambridge University Press, 2020.
Gregg, Heather S. "The Human Domain and Influence Operations in the 21st Century." *Special Operations Journal* 2, no. 2 (2016): 92–105.
Grove, Nicole Sunday. "Weapons of Mass Participation: Social Media, Violence Entrepreneurs, and the Politics of Crowdfunding for War." *European Journal of International Relations* 25, no. 1 (2019): 86–107.
Guevara, Ernesto Che. *Congo Diary: The Story of Che Guevara's "Lost" Year in Africa*. Lancing, UK: Ocean Press, 2015.
Gurr, Ted Robert. *Why Men Rebel*. Princeton, NJ: Princeton University Press, 1970.
Guy, Dennis C. *The Weaponization of Quantum Physics: How Technology Is Transforming Warfare*. Newport, RI: Naval War College, 2018.
Havel, Václav. *The Power of the Powerless*. October 1978. International Center on Nonviolent Conflict, www.nonviolent-conflict.org/wp-content/uploads/1979/01/the-power-of-the-powerless.pdf.
Headrick, Daniel. *The Tools of Empire: Technology and European Imperialism in the Nineteenth Century*. New York: Oxford University Press, 1981.
Healy, Jason, ed. *A Fierce Domain: Conflict in Cyberspace, 1986 to 2012*. Arlington, VA: Cyber Conflict Studies Association, 2013.
Heinlein, Robert A. *Starship Troopers*. New York: G. P. Putnam's Sons, 1959.
Herbert, Ulrich. "Berlin: The Persecution of Jews and German Society." In *Civil Society and the Holocaust: International Perspectives on Resistance and Rescue*, edited by Anders Jerichow and Cecilie Felicia Stokholm Banke. New York: Humanity in Action Press, 2013.

Herbst, Jeffrey. *States and Power in Africa: Comparative Lessons in Authority and Control*. Princeton, NJ: Princeton University Press, 2014 [2000].
Hibbert, Christopher. *Redcoats and Rebels: The American Revolution Through British Eyes*. New York: W. W. Norton & Company, 2002.
Higgins, Eliot. *We Are Bellingcat: An Intelligence Agency for the People*. London: Bloomsbury, 2021.
Hironaka, Ann. *Neverending Wars: The International Community, Weak States, and the Perpetuation of Civil War*. Cambridge, MA: Harvard University Press, 2009.
Hoffman, Frank G. "Examining Complex Forms of Conflict." *PRISM* 7, no. 4 (2018): 30–47.
———. "Foresight into 21st Century Conflict: End of the Greatest Illusion?" *Philadelphia Papers*, no. 14 (September 2016).
———. "Future Threats and Strategic Thinking." *Infinity Journal* 4 (fall 2011): 17–21.
———. "Squaring Clausewitz's Trinity in the Age of Autonomous Weapons." *Orbis* 63, no. 1 (2019): 44–63.
Hoofnagle, Chris Jay, Bart van der Sloot, and Frederik Zuiderveen Borgesius. "The European Union General Data Protection Regulation: What It Is and What It Means." *Information & Communications Technology Law* 28, no. 1 (2019): 65–98.
Howard, Michael. "The Forgotten Dimensions of Strategy." *Foreign Affairs* 57, no. 5 (1979): 975–986.
Howard, Philip N. *Lie Machines: How to Save Democracy from Troll Armies, Deceitful Robots, Junk News Operations, and Political Operatives*. New Haven, CT: Yale University Press, 2020.
Huang, Yanzhong. *Toxic Politics: China's Environmental Health Crisis and Its Challenge to the Chinese State*. New York: Cambridge University Press, 2020.
Hughes, Geraint. "War in the Grey Zone: Historical Reflections and Contemporary Implications." *Survival* 62, no. 3 (2020): 131–158.
Huntington, Samuel P. "Dead Souls: The Denationalization of the American Elite." *National Interest*, no. 75 (spring 2004): 5–18.
———. *The Soldier and the State: The Theory and Politics of Civil-Military Relations*. Cambridge, MA: Belknap Press, 1957.
Ibrahim, Azeem. *The Rohingyas: Inside Myanmar's Genocide*. New York: Oxford University Press, 2018.
Ikenberry, John. *The State*. Minneapolis: University of Minnesota Press, 1989.
International Commission on Intervention and State Sovereignty (ICISS). *The Responsibility to Protect*. New York: IDRC Books, 2001.
Jablonsky, David. "The Persistence of Credibility: Interests, Threats and Planning for the Use of American Military Power." *Strategic Review* (spring 1996).
Jackson, Robert. *Quasi-states: Sovereignty, International Relations and the Third World*. New York: Cambridge University Press, 1990.
Janowitz, Morris. *The Professional Soldier: A Social and Political Portrait*. Glencoe, IL: The Free Press, 1960.
Jayamaha, Buddhika. *Rebels, Inside and Out: Battlespaces in 21st Century Civil War*. Evanston, IL: PhD Dissertation, Northwestern University, 2018.
Jayamaha, Buddhika, and Jahara Matisek. "Social Media Warriors: Leveraging a New Battlespace." *Parameters* 48, no. 4 (2019): 11–24.
Johnson-Freese, Joan. *Space Warfare in the 21st Century: Arming the Heavens*. New York: Routledge, 2016.
Jomini, Le Baron de. *Précis de l'art de la guerre: Des principales combinaisons de la stratégie, de la grande tactique et de la politique militaire*. Brussels: Meline, Cans et Copagnie, 1838.

Jones, Hugh W. A. *Expanding the Battlefield: An Important Fundamental of Army-Multi-Domain Operations*. Carlisle Barracks, PA: Army War College Fellows, Strategy Research Project, 2021.
Jones, Bruce D., and Stephen John Stedman. "Civil Wars and the Post–Cold War International Order." *Daedalus* 146, no. 4 (2017): 33–44.
Jünger, Ernst. *Storm of Steel*. Translated by Michael Hofmann. New York: Penguin Classics, 2004 [1920].
Kaldor, Mary. *New and Old Wars: Organized Violence in a Global Era*. Hoboken, NJ: John Wiley & Sons, 2013 [1999].
Kallberg, Jan. "Supremacy by Accelerated Warfare Through the Comprehension Barrier and Beyond: Reaching the Zero Domain and Cyberspace Singularity." *Cyber Defense Review* 3, no. 3 (2018): 137–142.
Kalyvas, Stathis N. *The Logic of Violence in Civil War*. New York: Cambridge University Press, 2006.
Kalyvas, Stathis, and Laia Balcells. "International System and Technologies of Rebellion: How the End of the Cold War Shaped Internal Conflict." *American Political Science Review* 104, no. 3 (August 2010): 415–429.
Kane, John, and Haig Patapan. *The Democratic Leader: How Democracy Defines, Empowers and Limits Its Leaders*. New York: Oxford University Press, 2012.
Kaplan, Robert D. *The Coming Anarchy: Shattering the Dreams of the Post Cold War*. New York: Random House, 2002 [1994].
Karlin, Mara E. *Building Militaries in Fragile States: Challenges for the United States*. Philadelphia: University of Pennsylvania Press, 2018.
Kawczynski, Daniel. *Seeking Gaddafi: Libya, the West and the Arab Spring*. London: Biteback, 2011.
Kennan, George. "Policy Planning Memorandum," May 4, 1948, National Archives and Records Administration, RG 273, Records of the National Security Council, NSC 10/2.
Khong, Yuen Foong. *Analogies at War: Korea, Munich, Dien Bien Phu, and the Vietnam Decisions of 1965*. Princeton, NJ: Princeton University Press, 1992.
Kilcullen, David. *The Accidental Guerrilla: Fighting Small Wars in the Midst of a Big One*. New York: Oxford University Press, 2011.
———. *The Dragons and the Snakes: How the Rest Learned to Fight the West*. New York: Oxford University Press, 2020.
Kim, Bohyun. *Understanding Gamification* (Chicago: ALA TechSource, 2015).
King, Angus, and Mike Gallagher. *United States of America: Cyberspace Solarium Commission*. Arlington, VA: United States of America, Cyberspace Solarium Commission, March 2020.
Kissinger, Henry A. "The Congress of Vienna: A Reappraisal." *World Politics: A Quarterly Journal of International Relations* 8, no. 2 (1956): 264–280.
———. "Limited War: Conventional or Nuclear? A Reappraisal." *Daedalus* 89, no. 4 (1960): 800–817.
Klaits, Joseph, and Michael Haltzel, eds. *Global Ramifications of the French Revolution*. New York: Cambridge University Press, 2002.
Klar, Jeremy, ed. *The French Revolution, Napoleon, and the Republic: Liberté, Egalité, Fraternité*. New York: Rosen Publishing, 2016.
Klein, Maury. *A Call to Arms: Mobilizing America for World War II*. New York: Bloomsbury Publishing, 2013.
Knights, Michael. "Soleimani Is Dead: The Road Ahead for Iranian-Backed Militias in Iraq." *CTC Sentinel* 13 (January 2020): 1–10.
Koestler, Arthur. *Darkness at Noon*. Translated by Daphne Hardy. New York: Macmillian, 1941.

Kofman, Michael, and Matthew Rojansky. "What Kind of Victory for Russia in Syria?" *Military Review* (March–April 2018): 6–23.

Krasner, Stephen. *Sovereignty: Organized Hypocrisy*. Princeton, NJ: Princeton University Press, 1999.

Kreps, Sarah. *Social Media and International Relations*. New York: Cambridge University Press, 2020.

Krieg, Andreas, and Jean-Marc Rickli. *Surrogate Warfare: The Transformation of War in the Twenty-First Century*. Washington, DC: Georgetown University Press, 2019.

Kuehn, Kathleen M., and Leon A. Salter. "Assessing Digital Threats to Democracy, and Workable Solutions: A Review of the Recent Literature." *International Journal of Communication* 14 (2020): 2589–2610.

Kurzweil, Ray. *The Singularity Is Near*. New York: Penguin, 2005.

Lahr, M. M., et al. "Inter-group Violence Among Early Holocene Hunter-Gatherers of West Turkana, Kenya." *Nature* 529, no. 7586 (2016): 394–411.

Landau-Wells, Marika. "High Stakes and Low Bars: How International Recognition Shapes the Conduct of Civil Wars." *International Security* 43, no. 1 (2018): 100–137.

Lasswell, Harold D. "The Garrison State." *American Journal of Sociology* 46, no. 4 (1941): 455–468.

Levite, Ariel E., and Jonathan Shimshoni. "The Strategic Challenge of Society-centric Warfare." *Survival* 60, no. 6 (2018): 91–118.

Licklider, Roy. "The Consequences of Negotiated Settlements in Civil Wars, 1945–1993." *American Political Science Review* 89 (September 1995): 681–690.

Liddell Hart, Basil Henry. *A History of the World War, 1914–1918*. New York: Little, Brown. 1935.

———. *The Strategy of the Indirect Approach*. London: Faber and Faber, 1967.

Lind, William S., et al. "The Changing Face of War: Into the Fourth Generation." *Marine Corps Gazette* (October 1989): 22–27.

Lipset, David. *Gregory Bateson: The Legacy of a Scientist*. Englewood Cliffs, NJ: Prentice Hall, 1980.

Lipset, Seymour Martin, and Gary Marks. *It Didn't Happen Here: Why Socialism Failed in the United States*. New York: W. W. Norton & Company, 2000.

Locke, John. *Locke: Political Writings*. Edited by David Wootton. Indianapolis: Hackett Publishing, 1993 [1661].

Lord, C. G., L. Ross, and M. R. Lepper. "Biased Assimilation and Attitude Polarization: The Effects of Prior Theories on Subsequently Considered Evidence." *Journal of Personality and Social Psychology* 37 (1979): 2098–2109.

Lüthi, Lorenz M. "The Non-Aligned Movement and the Cold War, 1961–1973." *Journal of Cold War Studies* 18, no. 4 (2016): 98–147.

Luttwak, Edward N. *Strategy: The Logic of War and Peace*. Cambridge, MA: Harvard University Press, 2001 [1987].

Machiavelli, Niccolò. *The Prince*. Translated by G. Bull. London: Penguin Books, 1988 [1514].

Madsen, Jens Koed. *The Psychology of Micro-targeted Election Campaigns*. New York: Springer International Publishing, 2019.

Magnus, George. *Red Flags: Why Xi's China Is in Jeopardy*. New Haven, CT: Yale University Press, 2018.

Mahan, Alfred Thayer. *The Influence of Sea Power upon History, 1660–1783*. Boston: Little, Brown and Company, 1890. Available at Project Gutenberg, www.gutenberg.org/files/13529/13529-h/13529-h.htm.

Malcomson, Scott. *Splinternet: How Geopolitics and Commerce Are Fragmenting the World Wide Web*. New York: OR Books, 2016.

Mann, Michael. "The Autonomous Power of the State: Its Origins, Mechanisms and Results." *European Journal of Sociology / Archives Européennes de Sociologie / Europäisches Archiv für Soziologie* 25, no. 2 (1984): 185–213.
———. *Fascists*. New York: Cambridge University Press, 2004.
———. *The Sources of Social Power*. 4 vols. New York: Cambridge University Press, 1986.
Mansoor, Peter R., and Williamson Murray, eds. *The Culture of Military Organizations*. New York: Cambridge University Press, 2019.
Mastny, Vojtech, and Malcolm Byrne, eds. *A Cardboard Castle? An Inside History of the Warsaw Pact, 1955–1991*. New York: Central European University Press, 2005.
Matisek, Jahara. "The Crisis of American Military Assistance: Strategic Dithering and Fabergé Egg Armies." *Defense & Security Analysis* 34, no. 3 (2018): 267–290.
———. "Libya 2011: Hollow Victory in Low-Cost Air War." In *Air Power in the Age of Primacy: Contemporary Air Warfare Since the Cold War*, edited by Phil M. Haun, Colin F. Jackson, and Timothy P. Schultz. New York: Cambridge University Press, 2021.
———. "Shades of Gray Deterrence: Issues of Fighting in the Gray Zone." *Journal of Strategic Security* 10, no. 3 (2017): 1–26.
Matisek, Jahara, Travis Robison, and Buddhika Jayamaha. "Extending the American Century: Revisiting the Social Contract." *Georgetown Journal of International Affairs* 20, no. 1 (2019): 5–15.
Mattis, James N., and Frank Hoffman. "Future Warfare: The Rise of Hybrid Wars." *Proceedings Magazine* 132, no. 11 (November 2005): 30–32.
Mazarr, Michael J. *Mastering the Gray Zone: Understanding a Changing Era of Conflict*. Carlisle, PA: US Army War College, 2015.
Mazarr, Michael J., Ryan Bauer, et al. *The Emerging Risk of Virtual Societal Warfare: Social Manipulation in a Changing Information Environment*. Santa Monica, CA: RAND, 2019.
Mazarr, Michael J., Abigail Casey, et al. *Hostile Social Manipulation Present Realities and Emerging Trends*. Santa Monica, CA: RAND, 2019.
McCarthy, Kathleen. "From Cold War to Cultural Development: The International Cultural Activities of the Ford Foundation, 1950–1980." *Daedalus* 116, no. 1 (winter 1987): 93–117.
McKenzie, Mary M., and Peter H. Loedel, eds. *The Promise and Reality of European Security Cooperation: States, Interests, and Institutions*. Westport, CT: Praeger, 1998.
McMaster, H. R. *Battlegrounds: The Fight to Defend the Free World*. New York: Harper, 2020.
Mead, Walter Russell. *Special Providence: American Foreign Policy and How It Changed the World*. New York: Routledge, 2013 [2001].
Mearsheimer, John J. "Bound to Fail: The Rise and Fall of the Liberal International Order." *International Security* 43, no. 4 (2019): 7–50.
———. *Conventional Deterrence*. Ithaca, NY: Cornell University Press, 2017 [1983].
———. *The Great Delusion: Liberal Dreams and International Realities*. New Haven, CT: Yale University Press, 2018.
Mearsheimer, John J., and Stephen M. Walt. *The Israel Lobby and US Foreign Policy*. New York: Macmillan, 2007.
Melin, Elina. "China's Sharp Power Through TikTok: A Case Study of How China Can Use Sharp Power Through TikTok." Master's thesis, Linnaeus University, Sweden, 2021.
Melvern, Linda. *Conspiracy to Murder: The Rwandan Genocide*. New York: Verso, 2006.

Mencutek, Zeynep Sahin, and Bahar Baser. "Mobilizing Diasporas: Insights from Turkey's Attempts to Reach Turkish Citizens Abroad." *Journal of Balkan and Near Eastern Studies* 20, no. 1 (2018): 86–105.

Menkhaus, Ken. "Somalia: Unwanted Legacy, Unhappy Options." In *America's Challenges in the Greater Middle East*, edited by Shahram Akbarzadeh. New York: Palgrave Macmillan, 2011.

Meyer, Stephen A. "How the Threat (and the Coup) Collapsed: The Politicization of the Soviet Military." *International Security* 16, no. 3 (winter 1991–1992): 5–38.

Mikaberidze, Alexander. *The Napoleonic Wars: A Global History*. New York: Oxford University Press, 2020.

Miller, Edward S. 1991. *War Plan Orange: The U.S. Strategy to Defeat Japan, 1897–1945*. Annapolis: United States Naval Institute Press.

Millett, Allan R., and Williamson Murray, eds. *Military Effectiveness*. 3 vols. New York: Cambridge University Press, 1988.

Millett, Allan R., Williamson Murray, and Kenneth H. Watman. "The Effectiveness of Military Organizations." *International Security* 11, no. 1 (1986): 37–71.

Misra, Amalendu. *Politics of Civil Wars: Conflict, Intervention, and Resolution*. New York: Routledge, 2013.

Mölder, Holger, and Vladimir Sazonov. "Information Warfare as the Hobbesian Concept of Modern Times: The Principles, Techniques, and Tools of Russian Information Operations in the Donbass." *Journal of Slavic Military Studies* 31, no. 3 (2018): 308–328.

Morgenthau, Hans. *Politics Among Nations: The Struggle for Power and Peace*. New York: Knopf, 1960 [1948].

Morgenthau, Hans, and Ethel Person. "The Roots of Narcissism." *Partisan Review* 45, no. 3 (1978): 337–347.

Mott, William H. *United States Military Assistance: An Empirical Perspective*. Portsmouth, NH: Greenwood Publishing Group, 2002.

Mueller, Karl, ed. *Precision and Purpose: Airpower in the Libyan Civil War*. Santa Monica, CA: RAND, 2015.

Mumford, Andrew. *Proxy Warfare*. Cambridge, UK: Polity, 2013.

Munro, Geoffrey D., et al. "Biased Assimilation of Sociopolitical Arguments: Evaluating the 1996 US Presidential Debate." *Basic and Applied Social Psychology* 24, no. 1 (2002): 15–26.

Neiberg, Michael S. *The Treaty of Versailles: A Concise History*. New York: Oxford University Press, 2017.

Newman, Edward. *Understanding Civil Wars: Continuity and Change in Intrastate Conflict*. New York: Routledge, 2014.

———. "The Violence of Statebuilding in Historical Perspective: Implications for Peacebuilding." *Peacebuilding* 1, no. 1 (2013): 141–157.

North, Douglass C., John Joseph Wallis, and Barry R. Weingast. "Violence and the Rise of Open-Access Orders." *Journal of Democracy* 20, no. 1 (2009): 55–68.

Nowrasteh, Alex. "Espionage, Espionage-Related Crimes, and Immigration: A Risk Analysis, 1990–2019." Cato Institute, February 9, 2021, www.cato.org/publications/policy-analysis/espionage-espionage-related-crimes-immigration-risk-analysis-1990-2019.

O'Connell, Robert L. *Of Arms and Men: A History of War, Weapons, and Aggression*. New York: Oxford University Press, 1990.

O'Dea, S. "Global Smartphone Sales to End Users, 2007–2020." Statistica, March 31, 2021, www.statista.com/statistics/263437/global-smartphone-sales-to-end-users-since-2007.

Olson-Lounsbery, Marie, and Frederic Pearson. *Civil Wars: Internal Struggles, Global Consequences*. Toronto: University of Toronto, 2009.
Omissi, David E. *Air Power and Colonial Control: The Royal Air Force, 1919–1939*. Manchester, UK: Manchester University Press, 1990.
Osgood, Robert. *Limited War: The Challenge to American Strategy*. Chicago: University of Chicago Press, 1957.
Osinga, Frans P. B. *Science, Strategy and War: The Strategic Theory of John Boyd*. New York: Routledge, 2007.
Ozawa, Sachiko, et al. "Modeling the Economic Burden of Adult Vaccine-Preventable Diseases in the United States." *Health Affairs* 35, no. 11 (2016): 2124–2132.
Paterson, Thomas, and Lauren Hanley. "Political Warfare in the Digital Age: Cyber Subversion, Information Operations and 'Deep Fakes.'" *Australian Journal of International Affairs* 74, no. 4 (2020): 439–454.
Patrikarakos, David. *War in 140 Characters: How Social Media Is Reshaping Conflict in the Twenty-First Century*. New York: Basic Books, 2017.
Peceny, Mark. *Democracy at the Point of Bayonets*. University Park, PA: Penn State University Press, 1999.
Perego, Elizabeth. "Clampdown and Blowback: How State Repression Has Radicalized Islamist Groups in Egypt." *Origins: Current Events in Historical Perspective* 7, no. 10 (July 2014).
Perlroth, Nicole. *This Is How They Tell Me the World Ends: The Cyberweapons Arms Race*. New York: Bloomsbury Publishing, 2021.
Petersen, Roger D. *Western Intervention in the Balkans: The Strategic Use of Emotion in Conflict*. New York: Cambridge University Press, 2011.
Pettersson, Therese, and Magnus Öberg. "Organized Violence, 1989–2019." *Journal of Peace Research* 57, no. 4 (2020).
Pitt, Roger. "Warfare and Hominid Brain Evolution." *Journal of Theoretical Biology* 72, no. 3 (1978): 551–575.
Politi, Mauro. *The Rome Statute of the International Criminal Court: A Challenge to Impunity*. New York: Routledge, 2017.
Pomerantsev, Peter. *This Is Not Propaganda: Adventures in the War Against Reality*. New York: PublicAffairs, 2019.
Portzer, Joshua M. "Democratic Backsliding in the U.S. and Why the Rest of the West Should Care." *Inquiries Journal* 12, no. 2 (2020).
Posen, Barry R. *Inadvertent Escalation: Conventional War and Nuclear Risks*. Ithaca, NY: Cornell University Press, 2013 [1991].
———. "Is NATO Decisively Outnumbered?" *International Security* 12, no. 4 (1988): 186–202.
Powell, Robert. *Nuclear Deterrence Theory: The Search for Credibility*. New York: Cambridge University Press, 1990.
Power, Samantha. *"A Problem from Hell": America and the Age of Genocide*. New York: Basic Books, 2002.
Prager, Dennis. *Still the Best Hope: Why the World Needs American Values to Triumph*. New York: Harper Collins, 2012.
Price, David H. *Anthropological Intelligence: The Deployment and Neglect of American Anthropology in the Second World War*. Durham, NC: Duke University Press, 2008.
———. "Gregory Bateson and the OSS: World War II and Bateson's Assessment of Applied Anthropology." *Human Organization* 57, no. 4 (1998): 379–384.
Przeworski, Adam. *Crisis of Democracy*. New York: Cambridge University Press, 2019.
Putnam, Robert D. *Bowling Alone: The Collapse and Revival of American Community*. New York: Simon & Schuster, 2000.

Putnam, Robert D., and Shaylyn Romney Garrett. *The Upswing: How America Came Together a Century Ago and How We Can Do It Again.* New York: Simon & Schuster, 2020.
Qiang, Xiao. "The Road to Digital Unfreedom: President Xi's Surveillance State." *Journal of Democracy* 30, no. 1 (2019): 53–67.
Qiao Liang and Wang Xiangsui. *Unrestricted Warfare.* Beijing: PLA Literature and Arts Publishing House, 1999.
Qutb, Sayyid. *Ma'ālim fī al-ṭarīq.* 6th ed. Cairo: Dar al-Shuruq, 1979 [1964].
Reddaway, W. F. *Frederick the Great and the Rise of Prussia.* New York: Haskell House, 1969 [1905].
Reeder, Joe R., and Robert E. Barnsby. "A Legal Framework for Enhancing Cybersecurity Through Public-Private Partnership." *Cyber Defense Review* 5, no. 3 (2020): 31–44.
Regan, Patrick M. *Civil Wars and Foreign Powers: Outside Intervention in Intrastate Conflict.* Ann Arbor: University of Michigan Press, 2002.
Reno, William. *Warfare in Independent Africa.* New York: Cambridge University Press, 2011.
Reno, William, and Jahara Matisek. "A New Era of Insurgent Recruitment: Have 'New' Civil Wars Changed the Dynamic?" *Civil Wars* 20, no. 3 (2018): 358–378.
Rhodes, Benjamin D. *United States Foreign Policy in the Interwar Period, 1918–1941: The Golden Age of American Diplomatic and Military Complacency.* Westport, CT: Praeger, 2001.
Richards, Imogen. "'Flexible' Capital Accumulation in Islamic State Social Media," *Critical Studies on Terrorism* 9, no. 2 (2016): 205–225.
Rid, Thomas. *Active Measures: The Secret History of Disinformation and Political Warfare.* New York: Farrar, Straus and Giroux, 2020.
———. *Cyber War Will Not Take Place.* New York: Oxford University Press, 2013.
Robinson, Colin D., and Jahara Matisek. "Assistance to Locally Appropriate Military Forces in Southern Somalia: Bypassing Mogadishu for Local Legitimacy." *RUSI Journal* 165, no. 4 (2020): 68–78.
Robinson, Linda, et al. *Modern Political Warfare: Current Practices and Possible Responses.* Santa Monica, CA: RAND Corporation, 2018.
Rosen, Stephen Peter. "Military Effectiveness: Why Society Matters." *International Security* 19, no. 4 (1995): 5–31.
———. *Winning the Next War: Innovation and the Modern Military.* Ithaca, NY: Cornell University Press, 1994.
Rubenson, Sven. *King of Kings: Tewodros of Ethiopia.* Addis Ababa: Haile Selassie I University Press, 1966.
Rynning, Sten. *NATO in Afghanistan: The Liberal Disconnect.* Palo Alto, CA: Stanford University Press, 2012.
Saideman, Stephen M., and Marie-Joëlle J. Zahar, eds. *Intra-state Conflict, Governments and Security: Dilemmas of Deterrence and Assurance.* New York: Routledge, 2008.
Sarotte, Mary Elise. "Spying Not Only on Strangers: Documenting Stasi Involvement in Cold War German-German Negotiations." *Intelligence and National Security* 11, no. 4 (1996): 765–779.
Sayle, Timothy Andrews. *Enduring Alliance: A History of NATO and the Postwar Global Order.* Ithaca, NY: Cornell University Press, 2019.
Schelling, Thomas C. *Arms and Influence.* New Haven, CT: Yale University Press, 1966.
———. *The Strategy of Conflict.* Cambridge, MA: Harvard University Press, 1980 [1960].

Schumpeter, Joseph. *Capitalism, Socialism and Democracy*. New York: Routledge, 2003 [1943].
Scoggins, Suzanne E. "Propaganda and the Police: The Softer Side of State Control in China." *Europe-Asia Studies* (December 14, 2020): 1–21.
Sharansky, Nathan. *Fear No Evil*. New York: Random House, 1988.
Singer, Peter Warren, and Emerson T. Brooking. *LikeWar: The Weaponization of Social Media*. New York: Houghton Mifflin Harcourt, 2018.
Skocpol, Theda. "Civil Society in the United States." In *The Oxford Handbook of Civil Society*, edited by Michael Edwards. New York: Oxford University Press, 2012.
Slaughter, Anne-Marie, and Lee Feinstein. "A Duty to Prevent." *Foreign Affairs* 83, no. 1 (2004): 136–150.
Sloan, Stanley R. *NATO, the European Union, and the Atlantic Community: The Transatlantic Bargain Challenged*. Lanham, MD: Rowman & Littlefield Publishers, 2005.
Snyder, Alvin. *Warriors of Disinformation: American Propaganda, Soviet Lies, and the Winning of the Cold War: An Insider's Account*. New York: Arcade Publishers, 1995.
Spruyt, Hendrik. *The Sovereign State and Its Competitors: An Analysis of Systems Change*. Princeton, NJ: Princeton University Press, 1996.
Stanley, Jason. *How Fascism Works: The Politics of Us and Them*. New York: Random House, 2018.
———. *How Propaganda Works*. Princeton, NJ: Princeton University Press, 2015.
Stanton, Jessica A. *Violence and Restraint in Civil War: Civilian Targeting in the Shadow of International Law*. New York: Cambridge University Press, 2016.
Stenge, Richard. *Information Wars: How We Lost the Global Battle Against Disinformation and What We Can Do About It*. New York: Atlantic Monthly Press, 2019.
Stern, Sheldon M. *The Week the World Stood Still: Inside the Secret Cuban Missile Crisis*. Palo Alto, CA: Stanford University Press, 2005.
Storr, Jim. *The Human Face of War*. New York: Continuum, 2009.
Strachan, Hew. *Carl von Clausewitz's On War: A Biography*. London: Atlantic Books, 2013.
Strachan, Hew, and Sibylle Scheipers, eds. *The Changing Character of War*. London: Oxford University Press, 2011.
Tannenwald, Nina. "The Nuclear Taboo: The United States and the Normative Basis of Nuclear Non-use." *International Organization* 53, no. 3 (1999): 433–468.
Themnér, Anders, ed. *Warlord Democrats in Africa: Ex-Military Leaders and Electoral Politics*. London: Zed Books, 2017.
Thomas, Daniel. *The Helsinki Effect: International Norms, Human Rights, and the Demise of Communism*. Princeton, NJ: Princeton University Press, 2001.
Thomson, Janice E. *Mercenaries, Pirates, and Sovereigns: State-Building and Extraterritorial Violence in Early Modern Europe*. Princeton, NJ: Princeton University Press, 1996.
Thucydides. *History of the Peloponnesian War*. London: Penguin Books, 1972 [400 BC].
Tilly, Charles C. *The Formation of National States in Western Europe*. Princeton, NJ: Princeton University Press, 1975.
Tocqueville, Alex de. *Democracy in America*. Translated by Henry Reeve. New York: George Dearborn & Co, 1838.
Tucker, Spencer C., et al., eds. *The Encyclopedia of World War I: A Political, Social, and Military History*. New York: ABC-CLIO, 2005.

Van Creveld, Martin. *Technology and War: From 2000 BC to the Present.* New York: Simon & Schuster, 2010 [1989].
Van Evera, Stephen. "The Cult of the Offensive and the Origins of the First World War." *International Security* 9, no. 1 (1984): 58–107.
Veebel, Viljar, and Illimar Ploom. "Are the Baltic States and NATO on the Right Path in Deterring Russia in the Baltic?" *Defense & Security Analysis* 35, no. 4 (2019): 406–422.
Verini, James. *They Will Have to Die Now: Mosul and the Fall of the Caliphate.* New York: W. W. Norton, 2019.
Wahl, Paul, and Don Toppel. *The Gatling Gun.* New York: Arco Publishing. 1971.
Walt, Stephen M. *The Hell of Good Intentions: America's Foreign Policy Elite and the Decline of US Primacy.* New York: Farrar, Straus and Giroux, 2018.
———. "US Grand Strategy After the Cold War: Can Realism Explain It? Should Realism Guide It?" *International Relations* 32, no. 1 (2018): 3–22.
Watts, Clint. *Messing with the Enemy: Surviving in a Social Media World of Hackers, Terrorists, Russians, and Fake News.* New York: Harper, 2018.
Weber, Eugen. *Peasants into Frenchmen: The Modernization of Rural France, 1870–1914.* Palo Alto, CA: Stanford University Press, 1976.
Wells, H. G. *The Shape of Things to Come.* London: Hutchenson, 1933.
Westad, Odd Arne. *The Global Cold War: Third World Interventions and the Making of Our Times.* New York: Cambridge University Press, 2005.
Wilson, Peter H. *The Thirty Years War: Europe's Tragedy.* Cambridge, MA: Harvard University Press, 2011.
Wimbush, S. Enders, and Elizabeth M. Portale, eds. *Russia in Decline.* Washington, DC: Jamestown Foundation, 2017.
Woods, Jeff R. *Black Struggle, Red Scare: Segregation and Anti-communism in the South, 1948–1968.* Baton Rouge, LA: LSU Press, 2003.
Woolley, Samuel C., and Philip N. Howard, eds. *Computational Propaganda: Political Parties, Politicians, and Political Manipulation on Social Media.* New York: Oxford University Press, 2018.
Wylie, J. C. *Military Strategy: A General Theory of Power Control.* New Brunswick, NJ: Rutgers University Press, 1967.
Young, Francis. "Educational Exchanges and the National Interest." *ACLS Newsletter* 20, no. 2 (1969).
Zakem, Vera, Megan K. McBride, and Kate Hammerberg. *Exploring the Utility of Memes for U.S. Government Influence Campaigns.* Arlington, VA: Center for Naval Analyses, April 2018.
Zuber, Terence. *Inventing the Schlieffen Plan: German War Planning, 1871–1914.* New York: Oxford University Press, 2002.
Zuboff, Shoshana. *The Age of Surveillance Capitalism: The Fight for a Human Future at the New Frontier of Power.* New York: PublicAffairs, 2018.

Index

Abyssinia, 80, 95*n*44
Acheson, Dean, 53
Afghanistan, 3; as backyard of empires, 80–84; Cold War interstitial warfare in, 81; ethnic demography in, 95*n*50; Karzai president in, 83; modernization in, 81–82; NATO in, 79–84; punitive action in, 81–82; state-building in, 82–83; Taliban control of, 83; ultraconservatism in, 81; US in, 78–84. *See also* Taliban
AI. *See* Artificial intelligence
Air power, 37–38, 40–41, 168
American Civil War, 2, 105–106
Ancien régime, 98
Anomie, 8, 16*n*31
Arab Spring, 89–90
Artificial intelligence (AI): automation, Big Data and, 151–155; as bots, 135
Asymmetric irregular conflict, 78–80, 86
Australia, 108, 130–131, 134
Automation: AI, Big Data and, 151–155; as the present, 153–155

B-1 Lancer (aircraft), 94*n*7
Ba'athist state, 85–86
Balkans, 166; peacekeepers in, 82; state failure in, 75–77. *See also* Yugoslav Wars
Barre, Mohamed Siad, 70
Bateson, Gregory, 126

Battle of Mosul, 114*n*16
Battlespace, 4; altering strategic boundaries in, 98–100; contemporary complexity of, 5–10; cross-domain synchronicity in, 146; human agency shaping of, 63; integrated and synchronous strategic environment in, 145–146; of Libya, 88–91, 91*fig*; new horizons in, 155–156; technology shaping, 88–89; transformation of, 142–144, 144*figs*; unipolar moment and, 64–67; violence decreasing in, 8–9. *See also* Operational environment
Bensouda, Fatou, 101
Berlin, Isaiah, 28, 125
Bin Laden, Usama (UBL), 82, 84
Bismarck, Otto von, 100
Black Hawk Down, 71
Bowling Alone (Putnam), 127
Brandt, Willy, 56
Brezhnev Doctrine, 49
Britain, 99; indirect strategy by, 40–42; tank technology by, 152, 156
Brodie, Bernard, 46, 155
Bush, George H. W., 77, 95*n*39

Cameron, David, 129
Carr, E. H., 160
Castle Bravo test, 46
CCP. *See* Chinese Communist Party
Censorship, 92, 170

Chairman of the Joint Chiefs of Staff (CJCS), 39, 58
China: Australia influenced by, 108, 130–131, 134; consultants for, 129; data theft by, 132–133, 153; economic opening by, 64; espionage law in, 131; GPS system of, 112; internet in, 112; lawfare used by, 105; New Zealand influenced by, 130–131, 134; non-LSCO used by, 9; South China Sea claimed by, 5–6, 107–108; technological advances in, 152; UN undermined by, 149; in World Trade Organization, 107–108
Chinese Communist Party (CCP): businesses controlled by, 130; citizens pressured by, 131; cyberspace limited by, 122–123; digital surveillance by, 124
Churchill, Winston, 159–160
Citizenship, 126, 130
Civil society, 16n38; organizations in, 125–126, 134; schismogenesis in, 134–136; social engineering in, 121–122; as undefended, 121–124; weaponizing of, 126–129, 149
Civil war, 3, 15n9; American, 2, 105–106; Cold War prolonging of, 52–53, 59; international aid in, 74; between patronage networks, 73; political realities ignored in, 71–72; state failure and, 70–75; Syrian, 91–92, 106–107
CJCS. *See* Chairman of the Joint Chiefs of Staff
Clausewitz, Carl von, 19–21, 49, 136
Clinton, Bill, 121
Cold War, 12–13; civil wars prolonged by, 52–53, 59; deterrence strategy in, 46, 51; geopolitics in, 51–53; grand strategy in, 165; interstitial warfare in, 49–51, 59, 81, 100–101; military organizational reform and innovations in, 57–59; political warfare in, 53–57, 60; proxy warfare in, 52–53; strategic paradox in, 46–49; structural rigidity of, 68. *See also* Soviet Union; Unipolar moment
Congo, 72–73
Congress of Vienna, 98–100
Conventional warfare, 15n5
Council of National Defense, 38

Counter-Enlightenment, 20
Covid-19, 149
Crimea, 5–6, 49, 101–102, 104. *See also* Ukraine
Crisis of Agadir, 159–160
Cyberspace, 6; as borderless, 123; CCP limiting of, 122–123; data in, 113, 119–120, 137n4; disinformation campaigns in, 8, 16n30; malicious attacks in, 7; social engineering in, 121–122; virtual reality in, 118–119; virtual selves within, 118–121
Cyberwarfare, 6, 14, 121

Data: automation, AI and, 151–155; in espionage, 120–121; privacy of, 119–120, 137n4; regulation on, 119–121; theft of, 7, 132–133, 138n44, 153; tracking of, 113; weaponizing perception through, 132–136
Datascape, 128
Davos Man, 133
Deterrence, 46, 51, 161–162
Digital surveillance, 123–124, 128–129
Diplomatic, informational/ideational, military, economic (DIME), 144*figs*, 145–146
Domain, 6–7, 10, 16n38; emergent, 100–101, 111–113; multidomain, 13, 142, 147–151, 165–169. *See also* Civil society; Cyberspace; DIME; Outer space
Douhet, Giulio, 37–38
Dunlap, Charles, 105

Eastern Bloc, 65–66, 70. *See also* Soviet Union
Educational Propaganda Department, 38, 41
Enlightenment, 19–20
Espionage, 14, 129; data collection in, 120–121; fears of, 117, 131–132; in Soviet Union, 57
EU. *See* European Union
Europe, 39–40, 65
European Union (EU), 76, 119, 164
European Union Force Bosnia and Herzegovina, 82

F-35 (aircraft), 153, 168
Flashman, Harry, 81, 95n46
Ford Foundation, 55

Foreign-agent law, 147–148
Frederick the Great, 19
"Free world," 8, 47, 53
French Revolutionary Wars, 98; geopolitics in, 33–34; ideas in, 34–35; military organization in, 35; regime in, 34; scientific knowledge in, 35
Friedman, Thomas, 120, 121
Fukuyama, Francis, 68–69, 121
Fulbright-Hays Act, 55

Gatling, Richard, 17
Gatling gun, 17
General Data Protection Regulation (GDPR), 119, 121
Geneva Convention Common Article 3, 102
Geopolitics: in Cold War, 51–53; contemporary, 161; exploitation of, 101–103; in French Revolutionary Wars, 33–34; geostrategic periphery in, 51–53; in GRINS framework, 25–26; as idealized, 106; in indirect warfare, 51–53; transformation in, 150; transnational justice movements in, 77–78, 93, 148; in unipolar moment, 64, 67; in WWI, 36–37; in WWII, 39–40
Georgia, 102–103
Germany: blitzkrieg operations by, 38, 152, 156; civil society organizations in, 125–126; unification of, 98. *See also* West Germany
Global Positioning System (GPS), 89, 96n63, 112, 113. *See also* Satellites
Global War on Terror, 79, 136, 166
Globalization, 64
Goldwater-Nichols Department of Defense Reorganization Act, 58, 162, 167
GPC. *See* Great power competition
GPS. *See* Global Positioning System
Grand strategy, 160–161; in Cold War, 165; crisis in, 162, 165; material level challenges in, 168; as transformative enterprise, 164–165. *See also* Military strategy
Gray-zone activities, 97
Great Enlargement, 104–105, 113, 114n8; Middle East in, 108–111; South China Sea in, 107–108; Syrian Civil War in, 106–107
Great power competition (GPC), 9

GRINS framework, 11, 21–22, 22*fig*; application of, 31–32, 32*fig*, 160. *See also* Geopolitics; Ideas; Military organization; Regime; Scientific knowledge
Groundhog Day Disease, 165–166
Guillaume, Günter, 56
Gulf War, 164, 166

Harvey, Paul, 122
Havel, Václav, 66
Helsinki Accords, 65–66
Helsinki Watch. *See* Human Rights Watch
Hezbollah, 93
Hoffman, Frank, 19
House Un-American Activities Committee, 117
Houthi, 109
Human domain, 16n38
Human Rights Watch, 66

ICC. *See* International Criminal Court
Ideas: in French Revolutionary Wars, 34–35; in GRINS framework, 27–29; institutionalization of, 27–28, 43n34, 65–66; interventionist notion in, 74; live and let live, 79; rhetorical appropriation in, 102; terrorism influencing, 84; in World War I, 37
Ideational foundations: shifting of, 93; in unipolar moment, 64–65; in warfare, 24–25
IGO. *See* intergovernmental organizations
Immigrant, 130–131
Indirect warfare: geopolitics in, 51–53; in WWII, 40, 41–42. *See also* Interstitial warfare; Proxy warfare
Influence of Sea Power upon History, 1660–1783 (Mahan), 39
Information operations, 145–146
Information warfare, 6, 54–55. *See also* Sociopolitical-information warfare
Insurgent, 3, 79, 86, 90–91. *See also* Jihadist
Intelligence, surveillance, reconnaissance (ISR), 163
Intergovernmental organizations (IGOs), 65
International Commission on Intervention and Sovereignty, 74

International Criminal Court (ICC), 64–65, 78; on Russia's annexation of Crimea, 101–102; US against, 148
International law, 54, 63–64
International Republican Institute, 148
International tribunals, 148
Internet, 89–90; fragmentation of, 112; information accessible through, 121; as sovereign, 123–124
Internet of Things, 7, 126–127
Interstitial tactics: emergent domains and, 111–113; in outer space, 112
Interstitial warfare, 12; challenges of, 50–51; in Cold War, 49–51, 59, 81, 100–101; contemporary, 161–162; domains in, 101; indirect approach of, 49–51
Iran: against containment, 109; Iraq as client-state of, 110; Iraq War benefiting, 86; JCPOA agreement by, 107–109; in Syrian Civil War, 106–107
Iranian Revolutionary Guard Corps (IRGC), 109
Iraq, 3; Ba'athist state in, 85–86; as client-state of Iran, 110; consociational political arrangement in, 86; ISIS control in, 92; jihadists in, 86, 91; as modern bureaucratic state, 86; patronage networks in, 85; Shia militias in, 109–110; US invasion of, 84–86, 108, 148; WMD threat in, 84–85
IRGC. *See* Iranian Revolutionary Guard Corps
Irregular warfare, 3, 15*n*9, 79, 107. *See also* Asymmetric irregular conflict; Symmetric irregular conflict
ISIS. *See* Islamic State in Iraq and Syria
Islam, 84–85
Islamic State in Iraq and Syria (ISIS): civilian human shields used by, 106; emergence of, 91; Iraq controlled by, 92; rise and fall of, 91–93; Syria controlled by, 91–92; technology used by, 92; US defeat of, 161
Ismay, Lord Hastings Lionel, 164
ISR. *See* Intelligence, surveillance, reconnaissance

Al Jazeera, 90
JCPOA. *See* Joint Comprehensive Plan of Action

Jihadist, 83, 86, 91. *See also* Insurgent; Islamic State in Iraq and Syria (ISIS)
Joint Comprehensive Plan of Action (JCPOA), 107, 108–109, 114*n*21
Joint Force concept, 58, 77; in multidomain operational environment, 166–167; solutions in, 162–164
Joint Professional Military Education institutions, 163
Jomini, Antoine-Henri, 19

Karzai, Hamid, 83
Kennan, George F., 53–54, 103
Kessler syndrome, 112
KGB, 55
Khan, Imran, 84
Kinetic energy, 21
Kissinger, Henry, 129
Koestler, Arthur, 59
Korean War, 32, 45, 48
Kosovo, 102, 105. *See also* Balkans
Kosovo Force, 82

Large-scale combat operations (LSCO), 9
Law of the Sea, 107
Lawfare, 14, 105
LeMay, Curtis, 45, 88–89
LEO. *See* Low-Earth orbit
Libel, 118, 137*n*2
Liberal democracy, 8; armed forces in, 145–146; citizenship in, 126; civil society in, 125–126; doctrinal gaps in, 142; espionage fears in, 131–132; immigrants in, 130–131; juridical boundaries in, 154; military service in, 169; pluralism in, 79; privacy standards in, 132; strategic failures of, 143; subversion by stealth in, 141; undermining of, 150–151; virtual reality in, 149. *See also* Open society
Liberal internationalism, 164–165
Liberalism, 64–65, 147
Liberia, 72
Libya, 3, 7; asymmetric irregular conflicts in, 79; battlespace of, 88–91, 91*fig*; intervention urged for, 75; no-fly zone in, 88; Qaddafi in, 87–88; R2P in, 87–88; US disengagement with, 91; as War of Choice, 87–88
Licklider, Roy, 71

Liddell Hart, B. H., 38, 51, 152
Limited war: in Cold War, 49–50; definition of, 48
"Little green men," 5, 104
Lobby groups, 125
Louis Napoleon, 2; on battlespace, 4–5; as emperor, 31, 99; on Frederick the Great, 19; as undisputed leader, 34
Low-Earth orbit (LEO), 112
LSCO. *See* Large-scale combat operations
Lumumba Friendship University, 56

Mahan, Alfred Thayer, 39, 42
Mann, Michael, 25–27
Maxim gun, 17
McCain, John, 168
Military information support operations (MISO), 103
Military organization: Cold War innovations in, 57–59; combatant commander in, 58, 162; component commander in, 162–163; in French Revolutionary Wars, 35; in multidomain operational environment, 169; nature of, 29–30; in World War I, 38
Military power, 21, 25
Military strategy, 1–2; in contemporary multidomain environment, 165–169; crisis in, 160, 162, 165; definition of, 160–161; deterrence in, 46, 51; diplomatic communications in, 47–48; direct engagement in, 51; doctrinal level in, 162–163, 166–167; future strategy in, 169–171; Groundhog Day Disease in, 165–166; history influencing, 2, 170–171; institutional level in, 162–163, 166–167; material level in, 162–163, 166–168; nuclear conflict prevention in, 46; in open-ended conflicts, 166; personnel level in, 162–163, 166–169. *See also* Grand strategy
Misinformation, 16n30, 156
MISO. *See* military information support operations
Mitchell, Brig. Gen. William ("Billy"), 37–38, 41
Mitrokhin, Vasili, 50–51
Moltke, Helmuth von, 100
Morgenthau, Hans, 68
Myanmar, 122

Napoleonic Wars, 31, 34–35. *See also* French Revolutionary Wars
National Defense Education Act, 55
National Endowment for Democracy, 148
National Rifle Association (NRA), 134
National Science Foundation Network (NSFNET), 120
National Security Council Paper 68, 46, 162
Nationalism, 2, 27, 37
NATO. *See* North Atlantic Treaty Organization
New Zealand, 130–131, 134
9 Poma, 50
Nongovernmental organizations (NGOs), 65, 147–148
Normandy invasion, 40
North Atlantic Treaty Organization (NATO), 13; in Afghanistan, 79–84; in asymmetric conflicts, 79; establishment of, 47, 164; tactical gains of, 141; Yugoslav Wars intervention by, 76–77
Northern Alliance, 83
NRA. *See* National Rifle Association
NSFNET. *See* National Science Foundation Network
Nuclear weapons, 12; Castle Bravo test of, 46; in Cold War, 46, 155; in contemporary geopolitics, 161; as credible threat, 47; destructive power of, 45–46; as first Offset Strategy, 163; indirect warfare with, 48; in Korean War, 45; strategic paradox of, 46–49; in unipolar moment, 67

Offset Strategy. *See* Second Offset Strategy
On War (Clausewitz), 19–20
Open society: disadvantage of, 57, 170; fundamental characteristic of, 117; spread of, 121; strategic failures of, 143. *See also* Liberal democracy
Operation Just Cause, 164
Operational environment, 143, 144*figs*; contemporary military strategy in, 165–169; information operations in, 145–146; integrated multidomain of, 147–151, 166–169. *See also* Domain
Organizations: in civil society, 125–126; individuals, sharp power and, 129–132

Ottoman Empire, 26, 65
Outer space, 6–7; as congested, competitive, contested, 111–112, 115n27; as public and private domain, 111, 115n26; satellites in, 112–113

Pakistan: as opportunistic, 83–84; Taliban supported by, 82–83; UBL supported by, 84
Paradoxical trinity, 20–21, 25
Patronage networks, 73, 85
People's Friendship University. *See* Lumumba Friendship University
Political warfare: active measures in, 55–56; definition of, 53–54; international law in, 54; misinformation in, 156; MISO in, 103; open society disadvantage in, 57; Russian use of, 103; Soviet tactics in, 55–57; US tactics in, 53–55. *See also* Propaganda
Powell Doctrine, 51
Power, Samantha, 75
Propaganda: AI bots in, 135; CCP use of, 123; Educational Propaganda Department, 38, 41; ISIS use of, 92. *See also* Political warfare
Proxy warfare, 60n7; in Cold War, 52–53; in interstitial warfare, 12; political recognition in, 54; surrogates in, 52. *See also* Indirect warfare
Prussia: German unification enterprise of, 99–100; interstitial tactics of, 98
Przeworski, Adam, 141
Putin, Vladimir, 66–67
Putnam, Robert, 127

Qaddafi, Muammar, 87–88, 129
QAnon, 154–155
Qatar, 90
Quantum computing, 151–152
Quds, 91, 110

R2P. *See* responsibility to protect
Reconnaissance by fire, 109
Red Scare, 130
Regime: in civil wars, 72; in French Revolutionary Wars, 34; in GRINS framework, 26–27; shaping strategy, 145
Religion, 27

Responsibility to protect (R2P), 74, 87–88
Rome Statute, 78, 101–102
Rosenberg, Julius and Ethel, 117
Russia: civil society influenced by, 134; Crimea annexed by, 5–6, 26, 49, 101–102; cyber domain leveraged by, 103; foreign-agent law in, 147–148; Georgia invaded by, 102–103; GPS systems of, 112; information warfare used by, 6; lawfare used by, 105; "little green men" from, 5, 104; non-LSCO used by, 9; in Syrian Civil War, 106–107; troll farms of, 134; Ukraine dismembered by, 104–105; war crimes by, 101–102
Rwanda, 73–74

Sarkozy, Nicolas, 129
Satellites, 89, 112–113, 115n29. *See also* Global Positioning System (GPS)
Schelling, Thomas, 47, 162
Schismogenesis: in civil society, 134–136; definition of, 126; as internally generated, 141–142; twenty-first-century form of, 126–127; in US politics, 133–134; on virtual avatars, 124; in virtual reality, 127–128
Scientific knowledge: battlespaces shaped by, 88; definition of, 27, 30; duality in, 151; in French Revolutionary Wars, 35; in GRINS framework, 30–31; widespread applied use of, 89; in World War I, 37–38; in World War II, 39–40. *See also* Nuclear weapons; Technology
Second Offset Strategy, 162–163, 167–168
September 11 attacks, 78–79
Shalikashvili, John M. D., 126
Sharansky, Nathan, 66
Sherman, William Tecumseh, 88–89
Shia, 109–110
Sierra Leone, 70, 72
Slaughter, Anne-Marie, 74–75
Smart devices, 89, 92, 119–120
Smartphones, 90–91, 91*fig*, 96n65
Social engineering, 121–122
Social media, 90; censorship in, 92, 170; Chinese restriction of, 122–123; conspiracies in, 154–155; Hezbollah use of, 93; ISIS use of, 92; social

engineering through, 121–122; US regulation on, 122
Social source: military organization as, 29; of military power, 21, 25–31
Social space, 122, 137n10
Sociopolitical-information warfare, 8–9, 127. *See also* Information warfare
Somalia: state-building in, 72, 94n19; UNOSOM in, 70; US intervention in, 71, 74
South China Sea, 5–6, 107–108
Sovereign internet, 123–124
Soviet Bloc. *See* Eastern Bloc
Soviet Union: active measures of, 55–56; Brezhnev Doctrine in, 49; Communist indoctrination in, 57; espionage in, 57; implosion of, 58–59; political warfare of, 55–57; self-determination in, 69; sphere of influence of, 47–58; subversion by, 56, 117–118; US stalemate with, 46, 49. *See also* Cold War; Eastern Bloc
Special Operations Forces, 78
Stabilization Force in Bosnia and Herzegovina. *See* European Union Force Bosnia and Herzegovina
State failure, 24; in Africa, 70–75; in the Balkans, 75–77
State-building: in Afghanistan, 82–83; in Somalia, 72
Strategic Arms Limitation Treaty, 94n7
Strategic environment, 145–146, 150
Strategic narcissism, 13; definition of, 68; errors of, 164–165; foreign policy established from, 170; in Iraq invasion, 85; in unipolar moment, 67–69, 75
Strategy in the Missile Age (Brodie), 46, 155
Subversion, 8; in cyberwarfare, 14, 121; in liberal democracy, 141; Red Scare on, 130; by Soviet Union, 117–118; in West Germany, 56
Super PAC, 133
Surrogates, 52
Switzerland, 80–81
Symmetric irregular conflict, 86
Syrian Civil War, 91–92, 106–107

Taliban: Afghanistan controlled by, 83; modernism opposed by, 81–82; support for, 82–84

Technology: advances in, 152, 154; automation, AI, and Big Data in, 151–155; battlespaces shaped by, 88–89; for defense, 152; Second Offset Strategy in, 163, 167; US competitive edge in, 151; war altered by, 31; in warfare, 17–18, 37–38; in World War I, 36–37. *See also* Scientific knowledge
Terms-of-service agreement, 119, 137n4
Terrorism, 84–85, 154. *See also* September 11 attacks
Tewodros II (Emperor), 80
Third World, 49
Thirty Years' War, 24
Title 50, 167
Title 10, 167
Trade sanctions, 108–109
Transnational institutions, 12, 66, 101–102
Transnational justice movements, 77–78, 93, 148
Truman, Harry S., 50
Trump, Donald, 107–108; against JCPOA, 109, 114n21; kill order by, 110; Twitter blocking of, 122
The Twenty Years' Crisis, 1919–1939 (Carr), 160

UBL. *See* Bin Laden, Usama
Ukraine, 5, 103–105. *See also* Crimea
UN. *See* United Nations
UN Security Council: on Covid-19, 149; Russia in, 101–102; US influence in, 70–71
UN Security Council Resolution 688, 71
Unconventional war, 3, 15n10. *See also* Irregular warfare
Unified Command Plan, 162
Unified Task Force, 70
Unipolar moment, 13; aggression identified in, 104; battlespaces and, 64–67; end of, 93, 101; geopolitics in, 64, 67; liberalism in, 64–65; strategic narcissism and warfare in, 67–69, 75; US combat primacy in, 67–68, 77, 164; Western juridical arrangements in, 63–64
United Kingdom, 152–153
United Nations (UN), 54, 70, 102, 149
United Nations Operation in Somalia (UNOSOM), 70

United States (US): combat primacy of, 67–68, 77, 164; as defender of "free world," 53; espionage by, 57; against ICC, 148; lawfare used against, 105; market makers in, 152; on operational environment, 143, 144*fig*; political considerations for, 170; political warfare of, 53–55; proxy support motivations of, 53; Second Offset Strategy of, 163, 167; September 11 attacks in, 78–79; social behavior changes in, 127; social media regulations in, 122; Somali state intervention by, 71, 74; Soviet Union stalemate with, 46, 49; strategic crisis of, 161–162; strategic narcissism of, 13, 67–69, 75, 85; strengths of, 156; structural vulnerabilities in, 104; tactical gains of, 141; technology used by, 151, 163; trade sanctions by, 108–109; Ukraine guarantees by, 104; UN Security Council influenced by, 70–71. *See also* "Free world"; Liberal democracy

United States Information Agency (USIA), 54–55, 121

UNOSOM. *See* United Nations Operation in Somalia

US Agency for International Development (USAID), 74

US Department of Defense, 152–153, 156

USAID. *See* US Agency for International Development

USIA. *See* United States Information Agency

Venona Project, 117

Vietnam War, 32, 168

Violence: decreasing of, 8–9; in domain, 111; effects created by, 1, 7; explicit and implicit uses of, 5–6; moral logic of, 87; strategic communication through, 149; war without, 9–10

Virtual reality: avatars in, 119, 124, 146; in cyberspace, 118–119; illusive intensity and persistence in, 127–128; in liberal democracy, 149; scalability in, 128–129; schismogenesis in, 127–128; states in, 132; tactical depth in, 127–128

Voice of America, 55, 121

War: casus belli of, 23; definition of, 1, 4, 21; geopolitical context of, 25–26; ideas and, 27–29; levels in, 19, 22–23; *longue durée* in, 23–25; nature, character, outcome in, 21–23, 22*fig*; nature of military organization in, 29–30; objectives and outcomes in, 24–25; paradoxical trinity in, 20–21, 25; regime type in, 26–27; as social enterprise, 21, 42*n*15; without violence, 9–10

War crimes, 101–102

War Plan Orange, 39

"War-doves," 85

Warfare, 2; battlespace contours and, 4–5; character of, 11, 18–19, 42*n*5, 160; conquest prohibited in, 24; GRINS in, 21–22, 22*fig*; human ingenuity in, 17; ideational foundations in, 24–25; nature of, 11, 18–19, 23–24, 113; technological advances in, 17–18, 37–38

Warsaw Pact, 47–48, 57, 163

Weapons of mass destruction (WMD), 84–85, 96*n*53

West Germany, 56

WMD. *See* Weapons of mass destruction

The World Crisis (Churchill), 159–160

The World Is Flat (Friedman), 120

World Trade Organization, 65, 107–108

World War I (WWI), 31–32; educational propaganda in, 38, 41; geopolitics in, 36–37; ideas in, 37; military organization in, 38; scientific knowledge in, 37–38; technological advances in, 36–37

World War II (WWII), 2, 31–32, 38; geopolitics in, 39–40; indirect strategy in, 40, 41–42; scientific knowledge in, 39–40; strategic bombing in, 40–41

Yugoslav Wars, 75–77. *See also* Balkans

About the Book

WAR IS CHANGING. THE CYBERSPHERE, CIVIL SOCIETY, OUTER SPACE... all are emerging as domains in which battles are fought. What drives this shift? How is it affecting the character and conduct of war? What are the implications for military strategy?

As they address these fundamental questions, Jahara Matisek and Buddhika Jayamaha show how today's civil society, technology, and military organization are dramatically transforming warfare—in a world in which war is at once everywhere and nowhere, and nearly everything can be weaponized.

Jahara Matisek is associate professor of military and strategic studies, research director of the Strategy and Warfare Center, and senior fellow in the Homeland Defense Institute at the US Air Force Academy. **Buddhika Jayamaha** is assistant professor of military and strategic studies at the US Air Force Academy.